Climate Change isn't Everything

Climate Change isn't Everything

Liberating Climate Politics from Alarmism

Mike Hulme

polity

First published in 2023 by Polity Press

Polity Press
65 Bridge Street
Cambridge CB2 1UR, UK

Polity Press
111 River Street
Hoboken, NJ 07030, USA

ISBN-13: 978-1-5095-5615-1
ISBN-13: 978-1-5095-5616-8(pb)

A catalogue record for this book is available from the British Library.

Library of Congress Control Number: 2022948540

Typeset in 11 on 14 pt Sabon
by Cheshire Typesetting Ltd, Cuddington, Cheshire
Printed and bound in Great Britain by TJ Books Ltd, Padstow, Cornwall

For further information on Polity, visit our website:
politybooks.com

Contents

Contents

Acknowledgements

The idea for this book slowly gestated during the years since 2018, which have witnessed a growing alarmist discourse surrounding climate change, fuelled by the rhetoric of 'ten more years' and a proliferation of ticking clocks. During these years, I published a few short essays and commentaries – in the academic literature as well as on my blog – which drew attention to some of the dangers of this alarmist rhetoric. My idea that what I was witnessing was the emergence of an ideology I call 'climatism' catalysed during 2021. Particularly helpful in this catalysis were stimulating discussions about many aspects of climate change conducted with my graduate reading group in Cambridge: David Durand-Delacre, Freddie Hartz, Maximilian Hepach, Anneleen Kenis, Noam Obermeister, Kari de Pryck and Tom Simpson. While thanking them for their challenges and provocations to my thinking, the line of argument I develop in this book is distinctly mine, not theirs. I also wish to acknowledge the perceptive, and mostly encouraging, remarks offered by three anonymous readers of, first, an outline proposal for the book, and then later a full draft version of the manuscript. My colleagues Shin

Asayama, Rob Bellamy and Anneleen Kenis also offered valuable commentaries on a draft manuscript, and for this I am grateful. At Polity Press, I have benefited from the active support and advice of my editor Jonathan Skerrett. Notwithstanding the above, all opinions and judgements expressed here are my own.

Mike Hulme
Cambridge, December 2022

Introduction

Civil War, Racist Tweets and Flood Devastation

Following Friday prayers in the city's mosques, public protests spilled onto the streets of Dara'a, a smallish city in southwest Syria. Sunni Muslims were protesting against the recent arrest and torture of schoolboys in the city, some as young as thirteen, perpetrated by President Assad's security agents. These events happened in 2011, on Friday, 18 March; in the weeks that followed, protests against Assad's regime – led by activists and public intellectuals – erupted in other cities around the country. The localized protests in Dara'a turned into a national uprising. Assad attempted to suppress the protests, often violently, but by September street-fighting between government troops and newly organized anti-government militias was a regular sight in several Syrian cities.[1]

This uprising seeded the Syrian civil war which, over the following twelve years, is estimated to have claimed half a million lives and led to untold suffering. Of Syria's total population of 21 million, around 6 million were displaced internally by the war and a similar number were forced by the conflict to flee the country. As the fighting intensified, the number of Syrian refugees seeking asylum in Europe swelled, peaking at more than 30,000 asylum applications per month during the summer and autumn of 2015. At the same time, large numbers of additional refugees were arriving in Europe from Kosovo, Afghanistan, Albania and Iraq.

The resulting European migrant crisis of 2015 created major political ruptures across the continent as governments and opposition politicians reacted in different ways to the crisis. Many pointed to climate change to attribute blame for the fighting and the subsequent flow of refugees arriving in southeast Europe. Media headlines shouted, 'Syria's brutal four-year war blamed on CLIMATE CHANGE', 'Climate change opened "Gates of Hell" in Syria' and 'Climate change led to rise of Syrian terror'. These headlines fuelled a narrative which became widespread during 2015, and that has remained in circulation ever since: that the Syrian civil war was caused, at least in part, by human-caused climate change. This narrative ran as follows. A multi-year drought during the years 2006 to 2009, affecting most severely the agricultural areas of northeast Syria, was blamed on climate change. The ensuing crop failures displaced large numbers of agricultural labourers who sought employment in the towns and cities of western and southwestern Syria. These incoming rural migrants were then identified as the driver of subsequent political unrest in urban centres such as Dara'a. The Syrian civil war was thus a consequence of climate change.

This 'Syrian war/climate change thesis' attracted much attention. Barack Obama, the then United States President, claimed in May 2015 that climate change-related drought 'helped fuel the early unrest in Syria, which descended into civil war', while a few months later the US Secretary of State, John Kerry, argued that 'it's not a coincidence that immediately prior to the civil war in Syria, the country experienced its worst drought on record'. In November 2015, the heir to the British throne, Prince Charles, announced that 'there is very good evidence indeed that one of the major reasons for this horror in Syria was a drought that lasted for five or six years'. Many other institutions and organizations argued similarly – the World Bank, Friends of the Earth, official governmental and intergovernmental reports, defence think-tanks, academics, activists, and commentators of various political persuasions.[2]

The claim that the Syrian civil war was triggered by climate change gained salience in the weeks leading up to the important international climate negotiations – COP21 – that took place in Paris in December that year. The claim circulated in news media, threaded its way through countless social media posts, and fuelled fears about 'tides of climate migrants' crossing borders. The European Commission President, Jean-Claude Juncker, identified climate change as one of the 'root causes' of the new migration, with others suggesting that the displaced Syrians arriving in Europe were 'climate migrants' or 'climate refugees'. Eighteen months later, in March 2017, former US Vice-President Al Gore even claimed that, as a consequence of these migrants, Britain's vote in June 2016 to leave the European Union – the Brexit referendum – was 'partly caused by climate change'.[3]

For its advocates, the Syrian war/climate change thesis was important for two reasons. It offered a climatic explanation for the events of 2015, suggesting that the triggering effects of climate change on human conflict and migration are already with us. But this thesis also presaged the future chaos that some believe will ensue as greenhouse gas emissions continue to rise. As President Obama argued in his acceptance speech for the Nobel Peace Prize in December 2009, climate change will cause 'more drought, more famine, more mass displacement, all of which will fuel more conflict for decades'. The common refrain amongst the military is that climate change is a 'threat multiplier'. For those who argue that the number of migrants arriving in Europe will inevitably rise as the planet warms, the Syrian war/climate change thesis appeared as a harbinger of such a future, lending extra credibility to warnings of future climate-driven political instability.

Human-caused climate change is claimed to inflame conflict in other ways. A recent study has suggested that the volume of hate speech appearing on Twitter – content judged to be racist or xenophobic – is influenced by outdoor air temperature.[4] The authors analysed daily temperature data together with more than 10 million racist tweets originating between 2012 and 2018, emanating from six different countries spanning several climate zones across Europe. Their data showed that the volume of racist tweets was inversely proportional to the temperature distribution. The number of racist tweets and 'likes' was lowest for daily average temperatures between 5°C and 11°C, a range sometimes referred to as a 'climate comfort zone'. However, as the local daily average air temperature fell below 5°C and, even more so, as it rose above 11°C, the frequency of racist content increased steeply. The authors offered no causal explanation for this finding, but they claimed that their result showed that 'the occurrence and acceptance of racist content online could increase' in the future because the climate of Europe will get warmer. Within the next thirty years the number of days outside this climate comfort zone will increase across parts of the continent, suggesting to these authors that rising temperatures would aggravate xenophobia and racism in social media.

Climate change is also believed to cause other forms of devastation beyond war, migration and racist speech. Over a few days in the middle of July 2021, a month's rainfall fell over the Eifel Mountains of western Germany, near the border with Belgium and Luxembourg. The small river Kyll, and others like it, became raging torrents, and several riverside towns in the border area of Germany and Belgium were inundated. More than 200 people were killed across the region, and billions of euros of damage caused, in what was deemed the worst flood disaster in western Europe for several decades. Scientists pointed to the fact that a warmer atmosphere due to greenhouse gases had made such intense rainfall more likely. Many politicians blamed climate change for the disaster. The German Chancellor, Angela Merkel, toured the affected region and claimed that the world had 'to move faster in the battle against climate change'.[5]

*

What connects a destructive civil war in Syria, a rising volume of racist tweets, and devastating flood damage in Germany? It is the idea of climate change as a causal agent of adverse events. In each of these three cases, climate change was offered as the pre-eminent or decisive factor that 'explained' the troubling social outcomes described. These causal narratives are deployed by leading public figures, advocates and campaigners to urge a more vigorous 'battle against climate change', to stop climate change 'within ten years'. The implication of this framing – even if not always explicitly stated – is that if one wishes to reduce the likelihood of wars, the rise of racist tweets, the misery of flood damage, then limiting the future rate of climate change would be an effective way of doing so.

It is a line of thinking and argumentation that is alluring and which indisputably has become more widespread in recent years. But it's a way of thinking that's also dangerous. It draws attention to one set of realities – those related to humanity's ongoing activities which are re-shaping the world's climates – and elevates them to a universal explanation for all the world's ills. It identifies changing physical climatic conditions, but isolates them from the many local and regional contextual factors that condition the way in which meteorological events interact with historical, social, political, cultural, economic and ecological systems. This way of thinking ignores – or at least very seriously underplays – another, more complex, set of realities which condition the way the world works and, indeed, which condition the nature of the world's changing climate and the impacts it has on ecosystems, societies and people.

So, to take in turn the examples offered above: Rather than being caused by a severe drought in northeast Syria attributable to human emissions of greenhouse gases, the civil war in Syria was much more deeply rooted in that country's long history of ethnic tensions and political grievances exacerbated by President Assad's economic and social policies. The reasons why people promote racist tweets have much more to do with the changing modes and cultural norms of digital communication and with political influences on xenophobic attitudes than they have to do with the temperature outside people's windows. The flooding in the valleys of Germany's Eifel Mountains and the damage this caused was as much a result of changing land use in the upper catchments of these rivers – the spread of farms, the removal of natural vegetation, the expansion of impermeable surfaces – as it was a result of very intense rainfall. Mere association – or in some cases statistical correlation – between a climatic event and a deleterious outcome is not evidence of primary or even significant causation.

Limiting the rate of climate change over the next thirty or more years is a desirable long-term policy goal, but it should not be mistaken for interventions to prevent wars, defuse racism or contain flooding. For these, and for many other examples, there are better and faster ways to alleviate deleterious outcomes than to wait thirty or more years for global warming to be arrested. If one is concerned about the rise of racist tweets, the most effective way of tackling this is unlikely to be doubling-down on carbon dioxide emissions today: the avoided warming – and hence any putative defusing of the impulse to tweet racism – will not be realized for at least thirty years.

The purpose of this book is to warn against the allure of blaming everything on climate. It is to suggest that 'stopping climate change' might not always be the best way of stopping things from getting worse. It argues instead for placing whatever explanatory power climate change has for influencing affairs in the world into a much wider context. It argues that climate change isn't the *only* thing matters; it may not always be the most *important*. And it argues that doing everything one can, at all costs, to 'stop climate change' might at times even be a *distraction* from doing the things that really do make a difference.

*

Just over ten years ago, in 2011, I published an article in *Osiris*, a leading history of science journal: 'Reducing the future to climate: a story of climate determinism and reductionism'.[6] There, I introduced the term 'climate reductionism' to describe a particular way of thinking about the future which had gained ground in previous years. Climate reductionism, so I argued, imagined the future solely through the predictions of climate science, as though climate alone will determine the human future. I pointed out the deficiencies and the dangers of this way of thinking. The article has become the most cited in the journal's forty-year history.

Now, more than a decade later, a new variant of climate reductionism has taken hold. A way of thinking has gained a following that reduces not only the future to climate, but the present also. Contemporary politics is being reduced to the pursuit of a single overarching goal: to achieve net-zero carbon emissions by a given date, whether 2030, 2050 or whenever. By elevating *this* objective of political action in the world above all others, by making all other political goals subservient to *this* one, a dangerously myopic view of political, social and ecological well-being is being created. Whereas ten years ago I was concerned about how climate reductionist thinking was limiting our imagination of the *future*, I am now concerned about how it is constraining the politics of the *present*.

Climate reductionism has turned into a fully fledged ideology, an ideology I call 'climatism'. Climatism grows out of climate reductionism, but is more pervasive and insidious. At the same time, it is also more subtle and harder to isolate. At its most extreme, climatism uses the idea of climate change to 'naturalize' the problems of the world. The problems facing the world – whether the triumph of the Taliban, the management of wildfires, Putin's war in Ukraine, the movement of people – all become 'climatized'.

This 'naturalization' of social outcomes is similar to how biological racial theory has been used in the past – and sometimes still is today. According to racist thinking, some people struggle academically 'because' they are black; others are good at mathematics 'because' they are east Asian. So also with climatism: Some countries' economies underperform 'because' they have tropical climates; others go to war 'because of' climate change; people move 'because of' climate change; some people 'like' racist tweets 'because' it is hot outside; floods happen 'because' the rain is heavy. The instinct in common between climatism and racism is a desire to reduce understanding of the complexities of the world (whether human difference or social-ecological well-being) to a partial and incomplete scientistic project (whether biological race theory or climate modelling).

There are of course differences between climatism and racism, as I will make absolutely clear. Not least is the reality of human-caused climatic change. To some extent, this scientifically well-established fact 'de-naturalizes' the idea of climate. The effect of human influences on the climate system means that our climate can no longer be understood as simply 'natural'. Climate has now to be understood as something which is – at least partly – human-shaped. The patterns of weather around the world are indeed different than they would be on a twin planet without human presence. This distinction between climate (as natural) and ongoing *changes* in climate (as largely human-caused) is subtle and hard to characterize. It is a distinction that is easily elided in popular thinking and political discourse.

And it leads to two mis-steps.

The first is that *all* meteorological events become understood as mere proxies for human agency, whether the ultimate source of that agency is nefarious (e.g. fossil-fuel interests) or more innocent (e.g. meat-eating consumers). Climate's remaining 'naturalness' gets lost. Thus all hurricanes and heatwaves, for example, become viewed as manifestations of the behaviours of fossil-fuel companies, colonialism, capitalism, Amazonian loggers, rich meat-eaters or frequent flyers, forgetting that hurricanes and heatwaves are a natural feature of the world's climates.

Which leads to the second mis-step. Rather than seeking to understand the politics of why the impacts of similar meteorological phenomena on social and ecological well-being are so different in different places, attention is directed solely to the politics of 'stopping climate change'. This is a dangerous reduction in the scope of the political. To take hurricanes again as the example: the most pressing questions raised by the tragedy of Hurricane Katrina's impact on New Orleans in August 2005 pertain to the politics of race, flood defence and urban planning, not to the politics of burning fossil carbon or cutting down tropical forests.

Now in case anyone misunderstands me at this point, let me be absolutely clear. Just because hurricanes and heatwaves are natural features of local climates does not mean human actions cannot alter their intensity and/or frequency. And just because the impacts of weather and climatic extremes are always mediated by local social, economic and political factors does not mean we should ignore the need to decarbonize our energy systems and to manage our forests and land in general more sustainably. By pointing out the ideology of climatism and its attendant dangers, as I do in this book, I am not dismissing the scientific evidence that human actions have already caused changes in climatic patterns, and will continue to do so. This evidence is crystal clear. Nor am I suggesting that efforts to mitigate climate change and to adapt to its effects are worthless or should be stopped. What I am doing in this book is arguing *against* the ideology of climatism with its narrow and reductionist field of view, and in *favour* of a more contextually sensitive, diverse and pragmatic approach to incorporating the challenges of climate change into everyday politics.

*

Let me now turn to offer a short outline of how the book is organized. Chapter 1 – 'From Climate to Climatism' – explains what climatism is. I make clear that climate matters for human affairs. Of course it matters. Climates shape the physical and cultural environments within which all human life is conducted. And I also make clear the indisputable fact that the world's climates are changing in ways now significantly influenced by past and present human activities. That these climatic changes are afoot presents new and challenging contexts for human and non-human life alike.

But belief in these physical realities does not by itself constitute climatism.

No. Climatism becomes an ideology when it is held that the dominant explanation of social, political and ecological phenomena is 'a change in the climate', when complex political, social and ethical challenges become framed narrowly in terms of a changing climate. It becomes an ideology when these beliefs are held not just by assorted individuals, but when significant groups of people hold them to be self-evident social truths. The ideology of climatism claims that arresting climate change is the supreme political challenge of our time and that everything else becomes subservient to this one goal.

I argue that climatism, like other '-isms', is an ideology, a body of doctrine, myth and belief that guides individuals, institutions, classes, cultures or social movements. But if climatism is an ideology, why is this a problem? Ideologies are not 'wrong' simply because they are ideologies. After all, ideologies would appear to be inevitable constructions of the human mind. They are systems of belief that seem necessary for animating and guiding purposeful collective human action in the world. Many '-isms' are pernicious, for example racism, sexism, chauvinism or terrorism. But many are not, or at least not necessarily so. Think of impressionism, veganism, revivalism or syncretism.

By calling-out the ideology of climatism I am naming a way of thinking, arguing and acting in the world for what it is – a distinctive doctrine which, whilst providing animating power for some forms of political action, can also be self-justifying, discriminatory and oppressive. Climatism – as with other ideologies – is like a pair of tinted spectacles which colours how we see and interpret the world. Indeed, climatism is now so pervasive and embedded in many areas of public life that it is hard to recognize the spectacles we might be wearing. To expose climatism in our thought, speech and actions – and even more to challenge it – is to risk being seen to be undermining the reality of a changing climate or to be questioning the importance of taking climate into account when developing public policy.

But such undermining or questioning is not my aim. Arguing against the ideology of climatism does not mean arguing against the importance of climatic change for human affairs. Even less does it challenge the fact that humans are changing the climate. However, by challenging climatism I seek to free contemporary politics from two things. First, from the oppression of a scientized and deterministic view of the relationship between climate and society. And, second, from a dangerously myopic view that reduces the condition of the future world to the fate of global temperature or to the atmospheric concentration of carbon dioxide and other greenhouse gases.

The ideology of climatism has been constructed, and gained salience, gradually over the past forty years. In this development, it has been aided most significantly by a series of historical moves in the scientific and social scientific study of climate change. Ten such moves are elaborated in Chapter 2, 'How Did Climatism Arise?' These include the scientization of the study of climate, the adoption of global temperature as an object of policy, attempts to differentiate between human-caused and natural weather, and the claim that there are narrowing temporal windows within which change must be enforced, after which it will be 'too late'. These epistemic developments have allowed an influential array of thinkers, advocates and activists to appropriate the science of climate change to do political work for them. This appropriation, I believe, is one of the hallmarks of climatism.

Identifying the scientific underpinnings of climatism leads me in Chapter 3 to ask the question, 'Are the Sciences Climatist?' Here, I point out the danger that the climate sciences may become distorted because of the ideology of climatism. At the very least, it is possible that covert and overt pressures tempt climate science to lend its weight to an overtly political agenda. Studying humans scientifically through the lens of 'essentialized races' would be called out as racism. Similarly, studying societies, regions and cultures scientifically through the lens of 'essentialized climates' – the idea that climates have independent physical form and material agency in society – can be called out as climatism. This has implications for the sciences. Resting one's political and/or ethical arguments about how one acts in the world on a scientistic platform can lead to the authority of the underlying science being defended at all costs for fear of showing any weakness against one's political adversaries. As I explain, this defensiveness is good neither for science nor for politics.

So why has climatism become so pervasive? In Chapter 4, 'Why is Climatism So Alluring?', I lay out a number of features of climatism that help explain its appeal and rise to prominence. First among these is that climatism offers a master-narrative about the present and future state of the world. It is one that is sufficiently all-embracing and elastic that it is used to offer a seeming explanation for nearly everything – from the loss of sleep, to rising divorce rates, to the decline of insect populations in Europe, to electricity grid failures in Texas. As a master-narrative, climatism has recognizable similarities with other appealing ideologies – such as nationalism, apocalypticism and historicism – through which 'the true state of things' can be revealed. All such master-narratives hold one thing in common: tenacious belief in, and defence of, 'special knowledge' can explain the existence of otherwise bewildering social, cultural and even political phenomena.

Having explained why climatism is alluring, Chapter 5 then answers the question 'Why is Climatism Dangerous?' I elaborate five dangers of this ideological stance. First, climatism is always in danger of veering towards environmental determinism, the belief that human societies are 'prisoners of geography', to cite Tim Marshall's eponymous 2015 book,[7] or, in this case, prisoners of climate. Yet the effects that climates have on societies and ecologies are always heavily conditioned by a complex array of social, cultural, political and historical factors. Mono-causal explanations are usually wrong and sometimes dangerous. Second, through its adoption of apparently naturalized deadlines by which certain abstract numerical targets *must* be achieved, climatism creates a dangerous discourse of scarcity. Under this discursive condition there is not enough time to reflect, deliberate or experiment. Everything in public life is conducted 'in a hurry' which often leads to poor decision-making.

This then leads to a third danger, namely that climatism depoliticizes climate change. It reduces public politics to the politics of net-zero emissions; 'the ends justify the means', the motto of all totalitarian projects. Which, relatedly, points to the fourth danger, the anti-democratic impulse lurking within climatism. As a totalizing ideology, climatism brooks no public dissent. It seeks to police the boundaries of what can and cannot legitimately be said about climate change, not just about climate science, but also about climate politics and policies. Finally, through its myopic outlook on the world, climatism frequently leads to perverse outcomes. Climatist policies, quite reasonably, seek to reduce the extent of the human influence on climate. But in doing so they frequently create new – and sometimes quite troubling – social, ecological and political dangers and inequities.

If climatism is a dangerous ideology, then what is the alternative? Chapter 6, 'If Not Climatism, Then What?', offers some antidotes to the worst excesses of climatism. These include recognizing the uncertainties embedded in all climate predictions; dismantling the fear of falling over imagined climatic 'cliff-edges'; being more humble about the limits of knowledge and about the ability of strategic planning to manage the complex contingencies of the future; recognizing the diversity of political values and preferences within and between different polities; and setting policy goals that have a direct bearing on social-ecological welfare rather than striving to control abstract and scientistic global proxies for such welfare outcomes. Taken together, these antidotes offer a more pragmatic approach to incorporating the challenges of climate change in human affairs. They better recognize the political and geopolitical realities of international negotiations and national decision-making than does holding to the ideology of climatism.

There is a danger that the argument in this book will be mischaracterized, so in Chapter 7, 'Some Objections', I conclude by responding to some possible criticisms. For example, I tackle the objections that climate science is not alarmist, but in fact errs on the side of least drama; that climate change *is* an existential risk, and should be tackled as the number one priority; that 'justice' is much more central to climatists' policies than I imply; that the ideology of capitalism needs to be confronted with a competing ideology; and that, in the end, I sound just like a climate denier.

I

From Climate to Climatism

How an Ideology is Made

The land of Afghanistan has had its fair share of conflicts and miseries over the past 200 years. In the nineteenth century it was subject to an imperial tussle between Britain and Russia – 'the Great Game' as it was called – and in the past half century it has been invaded by both the Soviet Union and the United States and its western allies. More recently, it has faced internal conflict with the rising to power of the Taliban – a militant Islamist and jihadist political movement – who ruled the country between 1996 and 2001 and then again since August 2021. Throughout these imperial invasions, geopolitical manoeuvres and jihadist uprisings, the climate of Afghanistan has remained resolutely 'dry continental', to use the jargon of climatologists. It possesses an arid to semi-arid climate, very hot in summer and very cold in winter. Drought is endemic, rainfall rarely reliable but always welcome, and farming precarious. Afghanistan's climate enables and sustains a particular way of life and a distinctive set of agricultural practices.

And yet since the Taliban reclaimed power in 2021 from the American-backed democratic government of Mohammad Ahmadzai, the climate of Afghanistan seems to have gained new political agency. In the eyes of some western commentators, recent 'changes' in Afghanistan's climate have 'strengthened' the Taliban and even 'helped' it regain power. As one journalist argued, 'One of the decisive factors behind the Taliban's sudden takeover of Afghanistan this summer [2021] has been hidden in plain sight – climate change.'[1] The argument is made that new agricultural precarity, induced by 'drought or flood-ravaged soil', has been used by the Taliban to sow resentment against the former US-backed government and to recruit supporters to its cause. Being paid up to $10 per day to fight for the Taliban appears a more attractive deal for some young men, it is argued, than continuing to extract a livelihood from the land in the face of climate's vicissitudes.

We see in this one example how (easy it seems to be) to move from climate to climatism; from recognizing a nation's distinctive physical climate to offering 'a change in the climate' as a decisive explanation for political change. It is an example that we find repeated in many other spheres of contemporary life.

From climate to climate change

Climate may be defined in purely physical terms, as for example by the scientists of the Intergovernmental Panel on Climate Change (IPCC). For them, climate results from 'the evolving interactions between the five major components of the Earth's system: the atmosphere, the hydrosphere, the cryosphere, the lithosphere and the biosphere'. Alternatively, climate may be understood phenomenologically as weather routinely experienced in a place, or understood culturally through the kind of oral traditions that survive among many Indigenous peoples. However one might define it, climate is real. Either way, the seasonal rhythms of atmospheric phenomena which characterize a place's climate are a precondition for life. Put differently, one cannot imagine life in a climate-less world. A world that offered no order or pattern to the physical conditions emanating from the atmosphere – phenomena we call 'weather' – would be one in which it would be difficult, if not almost impossible, to live securely.

And yet it is now indisputable that the world's physical climates are changing in ways heavily influenced by the collective weight of historical and ongoing human activities. Around the world the seasonal rhythms of weather are changing. Establishing this fact has been a significant accomplishment of cooperative international science over the past century and more.[2] It has also been recognized by some more intuitively, through vernacular knowledge and first-hand experience. These changes afoot present a new and challenging dynamic for both human and non-human life. Their implications confront all social, ecological and political systems in today's world.

But this book is concerned neither with climate nor with climate change, both of which I have written about at length elsewhere.[3] Rather, it is about naming, explaining and challenging a very specific pattern of thinking about the world which is becoming increasingly dominant, and about the nature of consequent human action in the world. This pattern of thinking I call *climatism*. I label climatism an *ideology*, a structured set of beliefs that interprets social and political worlds and that is used to guide human action in those worlds.

Simple belief in the physical realities of a changing climate does not in itself constitute climatism, however these beliefs in climate may be scientifically expressed or culturally mediated. Neither is the mere recognition of the processes of *climatization* at work in today's world – how matters of concern become linked to a changing climate – constitutive of climatism. No. The ideology of climatism reaches out further than this. Climatism is the settled belief that the dominant explanation of social, economic and ecological phenomena is 'a human-caused change in the climate'. It frames the complex political and ethical challenges confronting the world today first and foremost in terms of a changing climate.

Yet climatism is a pattern of thought which I believe carries significant dangers for social justice, political freedom and future prosperity. This chapter lays out how one gets from recognizing the importance of climate and believing in human-caused climate change to the ideology of climatism. It explains how it is possible to claim that 'climate change helped the Taliban to win'; in other words, how it is possible to move from recognizing the distinctive physical characteristics of Afghanistan's climate and its changes, to becoming convinced that the rise of the Taliban is a result of climate change.

From climate to climatization

Climatization is the process whereby issues that were formerly deemed largely or totally unrelated to climate start being analysed and understood predominantly through a climatic lens. Thus *diet* becomes climatized when food choices are made on the basis of their possible impact on climate; *sport* becomes climatized when decisions on when, where and how to play become conditioned on climatic considerations; or *high energy particle physics* becomes climatized when the location of a future power-hungry hadron collider is determined by its carbon footprint. The list is long and growing of environmental and cultural phenomena, matters of human choice and behaviour, or public policy issues that have become or are becoming climatized. Over the last few decades, questions about human mobility and conflict, about urban design and transport planning, about recreation and tourism, about human fertility and fashion – and many more questions besides – have all become climatized. Even whale conservation has become climatized. For example, it is claimed that returning the whale population to the pre-whaling levels of the eighteenth century would shave off about 0.05°C from future global warming. This is because such an increase in the numbers of whales would sequester in the ocean significant amounts of carbon dioxide drawn-down from the atmosphere.[4]

Some further examples illustrate this trend towards climatization. Take the world of military professionals. In 2019, the International Military Council for Climate and Security (IMCCS) was launched at a conference on planetary security held at The Hague. The IMCCS comprises a worldwide group of senior military leaders, security experts and security institutions dedicated to 'anticipating, analysing and addressing the security risks of a changing climate'. As declared in its mission statement, 'Climate change is driving unprecedented risks to the geostrategic landscape of the 21st century. Militaries have a responsibility to help prevent and prepare for these risks.' Military powers have become climatized by evaluating the impacts of climate change on military assets and installations and through the rhetoric of 'greening defence'. For example, the UK's Parliamentary Defence Committee now holds regular hearings on defence and climate change. Military powers have become climatized by framing climate change as a 'threat multiplier' and by using climate-induced conflicts to justify and mobilize 'humanitarian assistance and disaster relief' by the military. And there are calls for greenhouse gas emissions from military activities to be separately accounted and for the impacts of war on climate to be factored into military strategy.[5]

Indian international relations expert Dhanasree Jayaram offers a detailed account of how this move towards climatization is playing out within the Indian armed forces. For example, a 'Green Cell' has been established at India's Navy Headquarters to 'coordinate and monitor implementation of the "green initiatives" by all segments of the navy', while India's Territorial Army has raised eight battalions in recent years to form an Ecological Task Force. This Force has been tasked with afforesting severely degraded land on the grounds that, with its 'military-like work culture and commitment', it can do the job more efficiently and decisively than civilian authorities.[6]

In a related vein, global geopolitical security has also become climatized. Through a series of debates organized at the United Nations' Security Council (UNSC), beginning in 2007, climate change has become a dominant framing of international relations and security. Political scientist Lucile Maertens suggests that the drivers for such climatization are threefold: strategic, instrumental and symbolic. *Strategically*, some member states seek national advantage by sponsoring climate change debates in the UNSC – a venue where those states perceive themselves to hold more influence – rather than under the umbrella of the UN Framework Convention on Climate Change (UNFCCC). *Instrumentally*, the UNSC seeks to climatize security concerns to overcome the perceived failures of global climate governance. The argument is that the UNSC can do for climate what the UNFCCC has failed to do. *Symbolically*, climatization of the UNSC seeks to elevate this body's importance in world affairs and to present it as the primary locus for managing global 'emergencies'. The end result of this process of climatization is an expansion of the political salience and agency of the UN's Security Council in relation to other potential sites of climate governance.[7]

Then there is the case of environmental disasters. Stephen Grant and colleagues observed the process of climatization at work in Bangladesh in 2009, when disastrous events and degraded environmental conditions were explained by climate change. In their view, 'climatization' occurred when flooding due to land-falling tropical cyclones was explained simply as 'the result of climate change', or when saltwater intrusion into some of the coastal areas of Bangladesh was blamed on climate change.[8] There has been a similar climatization of wildfires. Although wildfires are a ubiquitous natural phenomenon, indeed an essential one for maintaining ecosystem health, in recent years wildfires have come to 'stand in' for climate change in the public imagination. But as former wildfire fighter turned academic Stephen Pyne shows in his book *The Pyrocene: How We Created an Age of Fire, and What Happens Next*, there is a lot more to understanding wildfires than climate change. Indeed, one of the reasons for the recent climatization of wildfires has been, ironically, the too vigorous suppression of natural wildfires in preceding decades.[9]

A very different example of climatization comes from the work of religious history. In *Climate, Catastrophe and Faith: How Changes in Climate Drive Religious Upheaval*, historian Philip Jenkins develops a case for how climatic 'shocks' and fluctuations over the past millennium have re-shaped global religious beliefs and practices of spirituality. Jenkins argues that climatic change has often been accompanied by religious change.

The 'warmth and prosperity of the High Middle Ages' shaped Catholic doctrine in Europe, he says, and the Little Ice Age of the early modern period reduced Christians to a minority group in the Islamic world and left them persecuted in China. The climatization of religious history, exemplified by Jenkins, offers simple explanations for complex social and cultural phenomena such as religious revivals, heresies, persecutions and migrations.[10]

As these examples suggest, there are different motives behind this climatizing of the world's past, present and future. Climatization may be driven by the desire to find simple explanations for complex phenomena – an instinctive human reflex. We see above how this desire might play out in the case of religious phenomena and fire risk, but it is also at work in the climatization of civil conflict, suicide rates, economic decline, global pandemics . . . and much more besides. The trend towards climatization might also be motivated by wanting to exploit the material and symbolic capital associated with climate change and which can help issue advocates secure their intended political goals. There are often specific resources to be mobilized – financial and political – if issues can be climatized. Presenting them as 'climate change-related' is often likely to gain greater media attention, and hence greater public and political attention, for the relevant cause.

Climatization is also an attractive strategy because it can deflect public attention away from the underlying causes of problems which may have very little to do with climate change. These may be causes that some political actors would rather remain hidden. In the case of the Bangladesh disasters mentioned above, the regional devastation caused by Cyclone Aila in 2009 was due largely to unmaintained embankments and to the rapid influx of people into the most vulnerable coastal areas. And the saltwater intrusion in these areas was due mostly to the damming of major rivers and the pumping of shallow saline water sources for irrigation. Contra the headline of *The Guardian* article at the time which described these events – '"We have seen the enemy": Bangladesh's war against climate change' – Grant and his colleagues caution against this casual climatization of disasters in Bangladesh. 'Climatization [is] used as a means to cover up negligence or bad management', they say, and there is a risk that 'by climatizing a disaster, key vulnerabilities may be overlooked'.[11] This displacing of blame is an important feature of climatism to which I will return in later chapters.

From climatization to climatism

The combination of a physically changing climate and the discursive and political processes of climatization give rise to what political scientists Stefan Aykut and Lucile Maertens describe as a 'climatic logic in the making' (see Table 1). Their 'climate logic' has four characteristics which increasingly frame the way in which the idea of climate change is now publicly imagined. First, climate and its changes are understood almost exclusively through the results of scientific investigation. Climate has become 'scientized'. Second, and closely associated, climate becomes understood only from a planetary perspective. Climate has become 'globalized'. These characteristics of scientization and globalization mean that climate change discourses typically express a view from nowhere and offer a universal global gaze.[12] I will explore some of the origins of this framing of climate change in Chapter 2.

The third characteristic of Aykut and Maertens' climate logic is the extended temporality of climate change debates. This expresses itself in the belief that long-term strategic planning for the planet is not just desirable but also possible. This ambition for planetary governance is a novel one for humanity. Finally, their climate logic is also solution-oriented; in other words, it holds that there are market- or technology-based solutions to climate change waiting to be enacted and delivered. One expression of such solutionism is the rise of 'carbon reductionism', the quantification and evaluation of all public policies in terms of their carbon footprints. Another example would be the putative desire to implement solar climate engineering techno-fixes to manipulate global temperature. Also note that Table 1 offers possible antidotes to these characteristics of this emergent climate logic. I will return to consider some of these further in Chapter 6.

But the characteristics summarized in Table 1 are more than expressions of a 'climate logic in the making'. I want to push this further and say that they provide some of the key elements that enable and sustain climatism as an *ideology*. I will explore some of the features of this ideology in what follows, but for now notice the movement in thought that allows climatism to emerge. It starts with the study of climate and its changes, and

Table 1 Four characteristics of a 'climate logic' in the making, with associated expressions and antidotes

Characteristic	*Expression*	*Antidotes*
Scientized	View from nowhere	Plural ways of knowing
Planetary perspective	Global gaze	Alternative globalities
Long-term temporality	Strategic planning	Participatory futuring
Solution-oriented	Carbon reductionism	Social transformations

Source: Adapted and extended from Aykut and Maertens, 'The climatization of global politics'[13]

extends into predicting the future condition of the world on the basis of future climate, what I call 'climate reductionism'. It then ends up facilitating the widespread climatization of environmental and cultural phenomena, and hence of public policy issues. The culmination of this process of climatization is the ideology of climatism.

Similar movements in thought can be observed in other domains of analysis, discourse and action which lead to, for example, the ideologies of scientism, racism and globalism. Thus, *scientization* is the process by which all observable realities are considered explainable by physical properties and material causes which can be revealed through scientific method. *Scientism* then becomes the ideology – the settled belief, an interpretative pattern of thought guiding human action – that all conceivable questions about the world can be answered by science. Or take *racialization*. This is the process by which groups of people and their identities are defined 'naturally' by their race. *Racism* then becomes the ideology – the settled belief – that such racialized groups are differently and uniquely endowed and that individuals belonging to such groups can be ranked, judged and treated accordingly. Or try *globalization*. Here, abstract concepts, collections of facts, and material networks and flows come to be interpreted and hence understood as global in nature and scale. *Globalism* then becomes the ideology – the settled belief – that the whole world is the only possible and proper sphere for intellectual, cultural, economic and political activity and influence.

And so, by analogy, *climatization* is the process by which increasing numbers of public and political matters of concern come to be understood as shaped by changes in climate. *Climatism* then becomes the ideology – the settled belief – that the dominant explanation of all social, economic and ecological phenomena is a human-caused change in the climate. Complex political, socio-ecological and ethical problems and challenges come to be regarded as only being solved or adequately dealt with by first arresting the human causes of changes in climate.

RETORT

What Climatism is *Not*

'Climatism' is not a word I have invented, although it has not previously been widely used. Only one other book, to my knowledge, has explicitly named and explored the idea of climatism, Steve Goreham's *Climatism! Science, Common Sense and the 21st Century's Hottest Topic*,[14] and it is important that I disambiguate the term early on. Goreham, an American libertarian, uses the term in an entirely different sense to me. 'Climatism', he says, 'is an ideology promoting the belief that man-made greenhouse gas emissions are destroying the Earth's climate'. Goreham is denying any human influence on the climate system and therefore seeks to erase entirely from political thought and action the challenges presented by a (humanly) changing climate. In contrast, I fully recognize how climates around the world are changing in response to human activities. Yet I argue that in taking climate change seriously, contemporary politics should not be reorganized so as to deliver the single goal of stopping climate change. Our different positions regarding the ideology of climatism are very different and should not be confused.

Climatism as ideology

At its simplest, an ideology is a set of ideas, beliefs and values held by a particular social group that influences the way members of that group behave. In this sense, all of us have ideologies. We all hold a set, or sets, of beliefs and values that help us interpret the world we encounter and which therefore influence our actions. As political scientist Michael Freeden explains, 'We simply cannot do without [ideologies] because we cannot act without making sense of the worlds we inhabit.'[15] The ideologies to which we adhere therefore impose different filters on the world we encounter, filters which enable us to interpret diverse facts and experiences that would otherwise be too confusing. Political facts – indeed any facts – do not speak for themselves. In order to establish their meaning and significance, facts always need interpreting. We know that climates are changing and the process of climatization has expanded the range and complexity of climate 'facts' we are confronted with. Climatism has therefore emerged as an ideology which seeks to organize and interpret those facts so as to guide action in the world. We can think of ideologies as thought-maps by which we navigate the political and social worlds in which we find ourselves.

According to Freeden,[16] there are four key functions of an ideology, namely:

- it exhibits a recurring pattern
- it is held by significant groups
- it competes with others over controlling plans for public policy
- and it does so with the aim of justifying or changing the political arrangements of a political community

We shall see in later chapters how climatism fits these *functional* criteria, but for now let us recognize some additional *features* of ideologies.

Ideologies are often associated with the English suffix '-ism', which simply means 'taking side with' or 'in imitation of'. It is often used to describe philosophies, theories or artistic movements, as in commun*ism*, scient*ism* or cub*ism*. The attachment of '-ism' to an idea is often suggestive of a master-narrative or a totalizing explanation, something to which I shall return in Chapter 4 where I examine the allure of climatism. There is also a relationship between ideology and *tradition* in the sense that both offer tenets, doctrines or principles that guide belief and action.[17] Ideologies and traditions emerge when the possibility of questioning these principles – the fundamental 'idols of the tribe' – becomes unthinkable. At such times, ideologies and traditions can turn into dogmas.

This is not to say that climatism is in some sense inherently wrong or false. The suffix '-ism' is politically and ethically neutral. Yet 'isms', and hence ideologies, are often treated with suspicion. Thus, the term 'ideologue' – referring to an adherent of an ideology who is seen as dogmatic and uncompromising – is often used pejoratively. And '-isms' are often used to call out the assumed superiority of one group over another, as in rac*ism*, sex*ism*, species*ism*, or else to identify practices of discrimination operating on the basis of some physical attribute, as in age*ism*, abl*ism* or weight*ism*.

The ideologies identified above are resisted or valorized to the extent that the interpreter either disputes or shares them. For example, the ideologies of feminism, socialism and nationalism are 'ambi-valent'; that is, their valency, their normative orientation, is ambiguous and depends on the eye of the beholder. Thus feminists, socialists and nationalists find their political opponents among anti-feminists, anti-socialists and anti-nationalists. Climatism might therefore be a very effective ideology; it might offer a powerful interpretative map by which to navigate the contemporary world. It certainly has its allure, as we shall see. But as with any ideology, adopting climatism as a guiding thought-map has consequences and dangers to which we should be alert and about which we should be willing to be challenged.

Freeden's account of ideologies also makes a distinction between what he calls 'thin' and 'thick' ones. Thin ideologies are those that, whilst they have a recognizable shape and endurance, are quite limited in their interpretative ambitions and scope. Nationalism might be such an example. Thick ideologies, on the other hand, such as Marxism, offer more comprehensive doctrines for explaining how the world works, or how it should work. I suggest that whilst climatism might have started off as a thin ideology, it has aspirations to become in Freeden's sense a thick ideology, a totalizing one. One of the reasons for this aspiration is that climatism is rooted in a set of epistemic claims about the world emanating from science – see Chapter 2 – and which therefore appear to demand universal assent. Thick ideologies squeeze out dissent and demand full allegiance. The ideology of climatism claims that arresting climate change is the supreme political challenge of our time and that everything else becomes subservient to this one goal. The veteran British environment journalist George Monbiot expresses it thus: 'Curtailing climate change must . . . become the project we put before all others. If we fail in this task, we fail in everything else.'[18]

Climatism is an ideology that has varying degrees of visibility. It is most explicit in new social movements such as Fridays for Future, the Sunrise Movement or Extinction Rebellion. More broadly, it pervades the new 'climate left' who frame their political agenda almost solely in terms of arresting climate change. But climatism has also crept into a more extensive range of businesses, charities, professions and public institutions, such as Amazon, Oxfam, the BBC and the World Bank. Indeed, climatism is now so often the default position in some areas of public life that it can be hard to recognize and describe it without appearing to undermine the reality of a changing climate or to question the importance of taking climate into account when developing public policy in a rapidly changing world.

A good example of this is the case of investment strategist Stuart Kirk. In May 2022, Kirk, the head of responsible investing in HSBC's asset management group, gave a short fifteen-minute talk on environmental and social corporate governance (ESG for short) at a finance conference organized by the *Financial Times*. His presentation was titled 'Why investors need not worry about climate risk' and he accused central bankers and policymakers of overstating the financial risks of climate change. 'Unsubstantiated, shrill, partisan, self-serving, apocalyptic warnings are always wrong', he said. The sorts of statements Kirk had in his sights were such as, 'Climate change will dwarf the cost of living pain' (Mark Carney, former governor of the Bank of England), 'Climate change is the top global risk for business' (the World Economic Forum), and 'Firms must treat climate risk just as they do other risks' (the Bank of England).

The talk was available online and some of his remarks quickly went viral on Twitter and brought vicious criticism. For example, Christiana Figueres, the former executive secretary of the UNFCCC, called Kirk's comments 'outrageous' and demanded that he be fired. HSBC were duly forced to suspend Kirk from his job while they investigated. With soundbites such as 'Who cares if Miami is six metres under water in 100 years?' and 'There's always some nut-job telling me about the end of the world', Kirk's flippancy didn't help his defence. Six weeks later, Kirk resigned his position, stating: 'Ironically given my job title, I have concluded that the bank's behaviour towards me since my speech at a *Financial Times* conference in May has made my position, well, unsustainable . . . Funny old world.'[19]

I use Kirk's example not to defend everything he said. (There is no doubt that climate change *does* carry risks for investors, but Kirk was also drawing attention to legitimate questions about how those risks are best evaluated and assimilated.[20]) My point is just this. The ideology of climatism is so deeply embedded in the public imagination and, here, in parts of the financial establishment, that even the appearance of saying the 'wrong thing' or speaking in the wrong way is sufficient ground for public cancellation and professional opprobrium. I will say more about the 'chilling effect' of climatism on public speech in Chapter 5.

An example: disasters and climatism

The ideology of climatism in practice can be illustrated schematically using the paradigmatic case of climate-related 'disasters', such as those associated with meteorological extremes like floods, typhoons, heatwaves, droughts, and so on. Without reference to any human-caused *change* in climate, there have generally always been two ways of viewing such disasters. In stylized terms, disasters may be understood as resulting either from the meteorological hazard (the overlapping area in the scheme on the left in Figure 1) or from the prevailing socio-political conditions of the locality where the hazard occurs (the overlapping area in the scheme on the right in Figure 1).

In the first case, disasters are regarded as 'natural' or, in the case of climatic hazards, as 'climatized'. Historically, these have become institutionalized in the West under the rubric 'Acts of God', and they are still referred in this way for some purposes. More broadly speaking, such hazards are often regarded as the out-working of fate. Either way, responsibility for disasters is located in the impersonal forces of nature.

In the second case, disasters are de-naturalized, or 'socialized'. The focus of causation is placed not on the climatic hazard, but on the pre-existing sensitivities, exposures and vulnerabilities of people and assets to the hazard. As critical geographer and hazards researcher Danish Mustafa explains, 'it is an established principle within hazards research that a physical event only becomes a hazard when it comes into contact with vulnerable populations'.[21] Far from being the result of fate, these social conditions are understood to be a result of unfolding historical, economic and cultural processes, processes all mediated through the exercise of political power. This case is powerfully illustrated by Mike Davis in his book *Late Victorian Holocausts: El Niño Famines and the Making of the Third World*. Davis shows how the famines of the late nineteenth century in British India should not be understood as caused by drought, but rather by the laissez-faire and Malthusian economic ideology of colonial governments.[22]

In reality however, most disasters are the result of some combination of both the direct meteorological hazard – how heavy *was* the rain? – and the underlying exposure or vulnerability – how well-maintained *were*

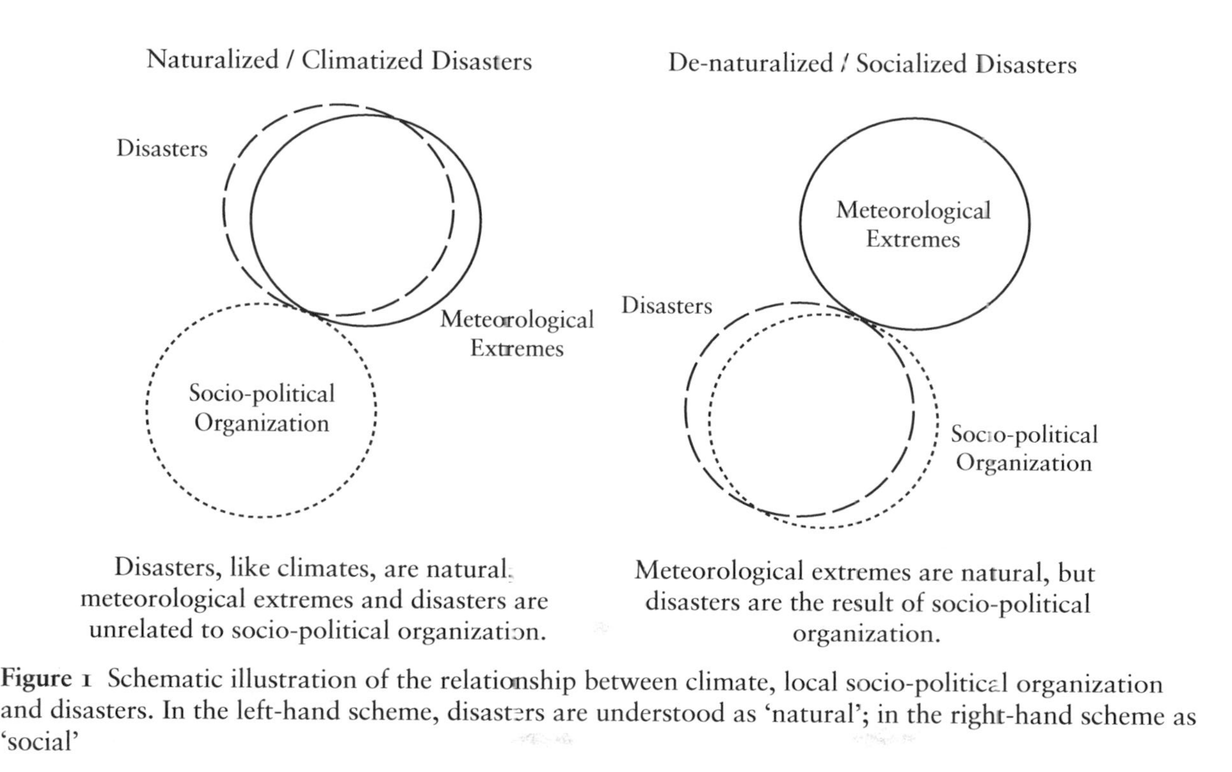

Figure 1 Schematic illustration of the relationship between climate, local socio-political organization and disasters. In the left-hand scheme, disasters are understood as 'natural'; in the right-hand scheme as 'social'

the storm sewers? For example, the extensive flooding in the provinces of Sindh and Balochistan in southern Pakistan in the summer of 2022 was as much a function of infrastructural and engineering choices as it was a result of an extraordinary monsoon. The relative importance of hazard, exposure and vulnerability for the disaster outcome varies widely from case to case and it is not always easy to ascertain the relative weights. Two different options are shown schematically in Figure 2. Political actors have a keen interest in how blame is apportioned. In *Natural Disasters: Acts of God or Acts of Man?*, a book written in 1984 in response to the then mounting famine in the Horn of Africa, authors Anders Wijkman and Lloyd Timberlake explored these difficulties of disaster attribution. Misreading the underlying causes of climate-related disasters, they argued, leads to misguided humanitarian interventions and policy goals.[23] Regardless of what happens in the future to the nature of climatic hazards, disaster outcomes will worsen if the socio-economic drivers of increased exposure and vulnerability are not tackled.

Thus far, these stylized cases have regarded climate as natural, as unaffected by human influences. A further spectrum of possibilities is created once it is recognized that the occurrence and/or the severity of meteorological hazards might be influenced by human-related emissions of greenhouse gases into the atmosphere and by other human-caused changes to atmospheric composition. Figure 3 illustrates schematically two ends of this spectrum, ranging from what I call lukewarmism to climatism.[24] The difference between these stylized cases is a matter of degree: by how much do the three domains of climate, society and disasters overlap? The difference between the lukewarmer and the climatist is revealed by the different sizes of the overlap between meteoro-

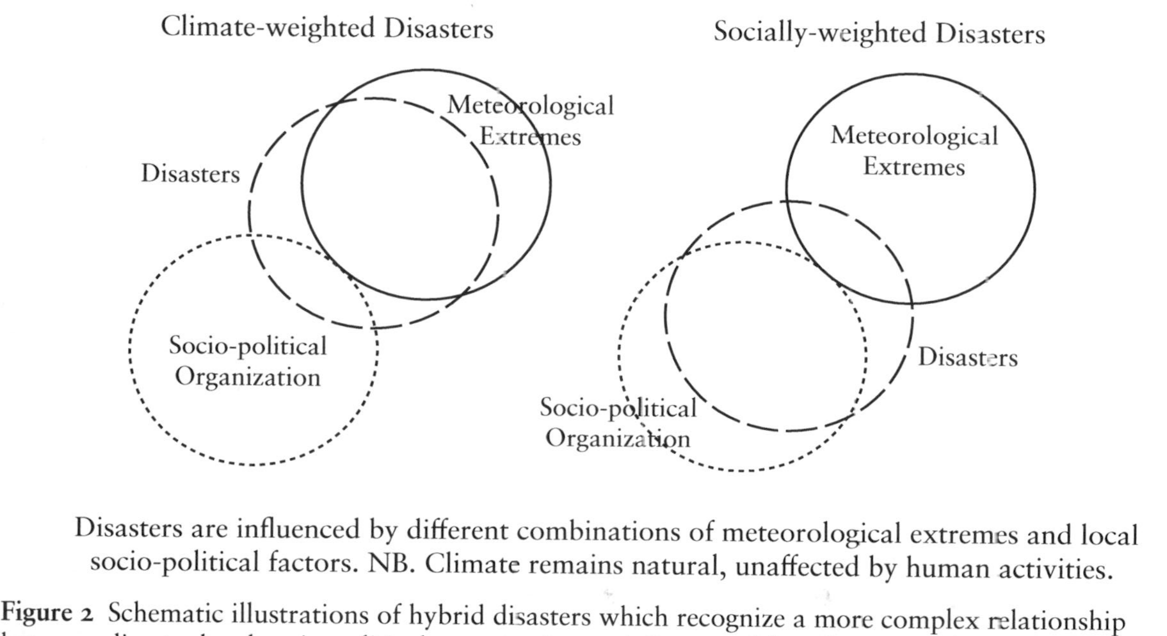

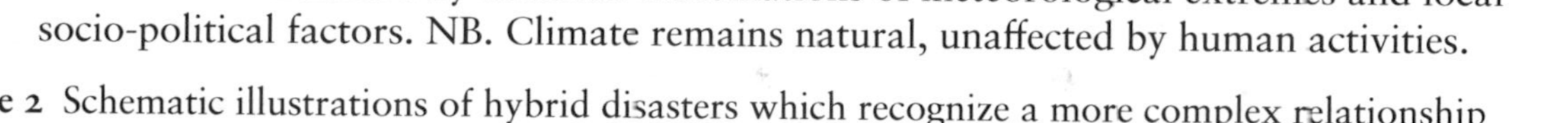

Figure 2 Schematic illustrations of hybrid disasters which recognize a more complex relationship between climate, local socio-political organization and disasters. The relative weighting of the meteorological extreme and the local socio-political organization in causing the disaster will vary from case to case and may not be easy to untangle

logical extremes and (in this case, global) socio-political organization.

But, as with the examples shown in Figure 2, the problem of attributing blame for the disaster remains. The lukewarmist and the climatist apply their respective interpretative filters – their ideologies – to answer the question. For the lukewarmist, human influences on climate, and hence on meteorological extremes, are believed to be quite modest and so disasters resulting from *natural* meteorological hazards will predominate, whether or not the impact of the hazard is heavily mediated by local socio-political factors. For the climatist, (nearly) all meteorologically related disasters appear as the result of human-changed climate; they have the fingerprint of human-caused climate change all over them, again whether or not the impact of the hazard is mediated by local socio-political factors.

Recent developments in climate science and modelling have sought to address this problem of attribution scientifically: 'How likely is it that this climatic extreme would have occurred in a world with no human influence?' But such scientific investigations under-determine the answer about the ultimate cause of disasters since they do not consider all relevant factors. Take the case of Hurricane Katrina devasting New Orleans in August 2005. Even *if* weather attribution science could establish that human influences made this particular hurricane 20 per cent, say, more intense, this wouldn't establish whether the ensuing deaths and the damage to the city's fabric were a function of the hurricane's intensity or of poor urban planning, flood defence or local policing. The disaster might well be the outcome of a natural meteorological hazard, intensified to some degree by human influences on the climate, interacting with local conditions mediated through the politics

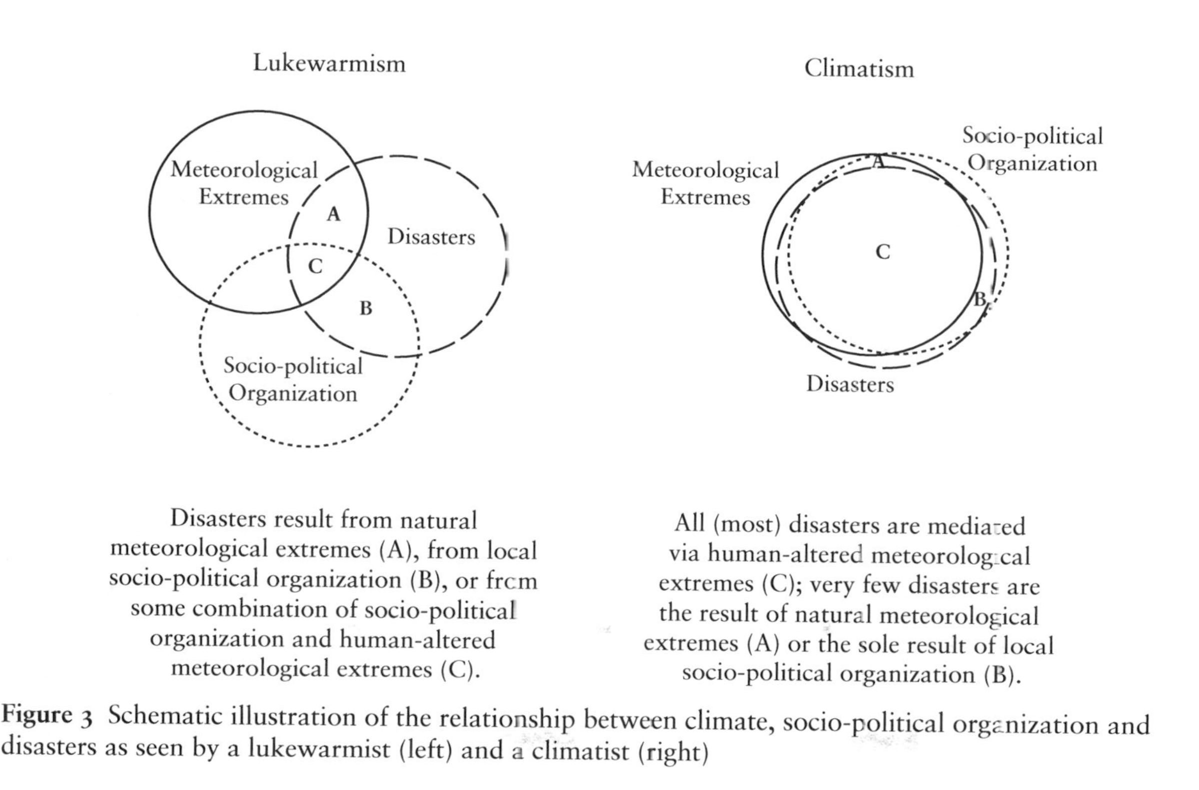

Figure 3 Schematic illustration of the relationship between climate, socio-political organization and disasters as seen by a lukewarmist (left) and a climatist (right)

of urban planning, zoning and flood defence. But how does one apportion blame for the disaster between these three elements? Weather attribution science does not provide the answer.

Summary

This chapter has outlined my argument that climatism should be regarded as a fully fledged ideology, a thin ideology certainly and, in Freeden's sense, in the process of becoming a thick one. I have shown how belief in a physically changing climate driven largely by human activities, combined with the discursive and political processes of climatization, gives rise to an emerging 'climate logic'. This logic increasingly structures the way in which the contemporary world and its future is being understood, and it leads to the ideology of climatism.

Climatism is a settled pattern of ideas, beliefs and values which holds that the dominant explanation of social, political and ecological phenomena is 'a changing climate'. To use the language introduced earlier, and in relation to climatic hazards, Acts of Man have replaced all Acts of God since climate itself is now an Act of Man; and all Acts of Man are now mediated though human-caused climate change. The implication of climatism is that complex social, ecological and political problems, and their associated ethical challenges, must be analysed – and solved – in terms of a changing climate. Preventing the *climate* from getting worse will prevent *everything* else from getting worse.

Before looking in detail at the allure of climatism (Chapter 4), and some of its attendant dangers (Chapter 5), there are two further steps to take. First, it is necessary to understand how this ideology has gained its ascendancy. Most ideologies gain their authority from legacies of patriarchy, from convictions about national identity, from historical analysis, from the interpretation of religious texts, and so on. But climatism gains its power from a set of universal epistemic claims about the world, emanating from the enterprise of science. It is founded on an increasingly intricate and interconnected set of scientific and social scientific claims established over the past forty years. It is these knowledge claims we must next investigate in Chapter 2. Furthermore, it will also be important to probe scientific practice itself to understand the extent to which the sciences and social sciences are themselves 'climatist'. This will be the task of Chapter 3.

2

How Did Climatism Arise?

Fetishizing Global Temperature

Following the Wall Street Crash of October 1929, the United States entered a period of economic turmoil and high unemployment – the Great Depression of the early 1930s. Unemployment rates peaked in March 1933 at around 20 per cent of the workforce, just as Franklin D. Roosevelt was inaugurated as the 32nd President of the United States. During his first 100 days in office, Roosevelt implemented a series of federal policies to provide relief for the unemployed and to stimulate the reform and recovery of the economy. Collectively, these executive orders and the ensuing legislation became known as the 'New Deal'. But Roosevelt's economic strategy was hampered by the lack of reliable measures of national economic performance. Policymakers at the time had no means of determining the extent of national economic contraction and nor, by implication, of economic recovery.

So, shortly after Roosevelt came to power, the US Senate commissioned a study on how to measure the output of the American economy. A thirty-three-year-old Russian émigré, Simon Kuznets, then on the staff of the National Bureau of Economic Research in Washington, led the study, which was published early in 1934. In his report for the US government, simply titled *National Income, 1929–1932*, Kuznets laid the groundwork for what was to become known as gross domestic product, or GDP, as a means to guide national economic planning as the country stumbled its way out of the Depression. Kuznets' idea was to combine all the produce, products and service goods generated by individuals, companies and the government into a single measure of economic activity. As he explained in the report:

> If all commodities produced and all personal services rendered during the year are added at their market value, and from the resulting total we subtract the value of that part of the nation's stock of goods which was expended . . . in producing this total, then the remainder constitutes the net product of the national economy during the year. It is referred to as national income produced, and may be defined briefly as that part of the economy's end-product which is attributable to the efforts of the individuals who comprise a nation.[1]

As formulated by Kuznets, GDP should rise in good times and fall in bad. Over the next few years, GDP began to replace well (or poorly) informed guesses about how much the economy was growing or contracting. It laid the ground for the new field of macroeconomic policy management. As we now know, GDP rapidly inscribed itself at the heart of modern societies, initially in the United States and other advanced economies in the post-war period, but then eventually in nearly all countries around the world. By the 1970s, GDP had become the measure by which all economic policies – and much else besides – were to be judged successful or not. Governments, political reputations, the public mood, the results of national elections . . . all rose or fell according to the performance of GDP.

About forty years after Kuznets' report to the Senate, another young American economist – William 'Bill' D. Nordhaus, the grandson of German immigrants – applied the idea of GDP in a novel way. With a grant from the US National Science Foundation, Nordhaus spent a year at the International Institute for Advanced Systems Analysis just outside Vienna in Austria. Here, in June 1975, he produced the first economic analysis of the cost of limiting global warming caused by emissions of carbon dioxide. His pioneering study asked two questions: 'How can we limit the concentration of atmospheric carbon dioxide to a reasonable level? And 'How much would a control path cost if it were implemented on an efficient basis?'[2]

To answer these questions, Nordhaus looked at the costs of replacing fossil-based energy fuels with non-carbon dioxide emitting energy technologies – for him these technologies were nuclear power and a range of renewable energies – to the degree necessary to 'control the climate'. But just as Kuznets found it necessary in the 1930s to create a new measure of economic activity, so Nordhaus had to rest his analysis on some measure of global climate which, in some way or other, represented human welfare. He fixed on the rather speculative notion of 'global temperature' as his control variable, arguing that a desirable world climate – what he called 'a stable climatic regime' – would vary by no more than ±1°C in global average temperature.[3]

Nordhaus was the first to combine analytically these two measures: GDP and global temperature, one measuring economic activity, the other the condition of global climate. He concluded his 1975 study thus:

> Clearly the control of carbon dioxide is not free – the medium control program has discounted costs of $37

> billion in 1970 prices. On the other hand, the cost as a fraction of world GNP [Gross National Product, a globally aggregated version of GDP] is likely to be insignificant, less than 0.2 per cent in the most stringent case.

Nordhaus subsequently went on to a prominent academic career as a world-leading climate economist, in 2018 receiving a joint share of the Nobel Memorial Prize in Economic Sciences. In its award testimony, the Royal Swedish Academy of Sciences specifically recognized Nordhaus's efforts to develop 'a quantitative model that describes the global interplay between the economy and the climate'.

This story of GDP and global temperature is important for two reasons. First, it draws attention to the equivalence by analogy of these respective measures of the health of the economy and the stability of climate. Both measures offer a target for policy – a single and seemingly easy-to-understand control variable – which can guide economic or climate policies respectively. Indeed, both measures have subsequently done so in profound ways: GDP to maximize economic wealth, global temperature to minimize the human destabilizing of climate. The second reason I tell this story is even more important. It is to alert us to the limitations of both of these measures for capturing everything that matters. In his original 1934 report to the US Senate, Kuznets himself noted this with respect to GDP:

> The above detailed classifications provide a fair description of the various groups of services which are included, at their market value, in the national income. *But they are far from an exhaustive account* of the possible contents and scope of the national income measurement. The boundaries of a 'nation' in 'national' income are still to be defined; and a number of other services . . . might also

> be considered a proper part of the national economy's end-product. [emphasis added].

In this and later chapters we shall see some of the equivalent limitations of global temperature as an index that captures all that matters about climate and welfare. The danger of quantitative indicators such as GDP and global temperature is that they gain an independent power of their own. They bedazzle users and advocates because they are believed to reveal more about the state of the world – whether about the economy or about the climate – than they actually do. Using a term from anthropology, both GDP and global temperature become 'fetishes' – that is, things to which the human imagination grants causal powers in the material world even though they are themselves inventions of the human mind. As fetishes, they can become dangerous. In his 2015 book, *The Little Big Number: How GDP Came to Rule the World and What to Do about It*, the economic historian Dirk Philipsen points this out:

> GDP has ballooned from a narrow economic tool into a global article of faith . . . this spells trouble. While economies and cultures measure their performance by it, GDP only measures output. It ignores central facts such as quality, costs, or purpose. Sustainability and quality of life are overlooked. Losses don't count. The world can no longer afford GDP rule – GDP ignores real development.[4]

What GDP and global temperature reveal about the economy and about climate may not be the things that matter the most.

How Did Climatism Arise?

Framing climate change

This chapter explains how the ideology of climatism arose – and how it did so gradually, almost imperceptibly, over the course of recent decades. Most significant in this explanation is a series of developments in the scientific and social scientific study of climatic change starting in the 1980s. One of these developments – perhaps the most important one – was the adoption of global temperature as an index deemed a suitable proxy measure of the state of the world's climate as it mattered to people and societies. Much as GDP came during the latter half of the twentieth century to define the health of economies, so global temperature has come more recently to define the health of the world's climate.

But the ideology of climatism rests on more than just the index of global temperature. What follows are brief descriptions of ten developments – I call them 'moves' – in the scientific and social scientific study of climatic change over the past forty years that have contributed to climate change being understood in ever narrower and more reductionist terms. The cumulative effect of these moves has enabled thinkers and analysts, advocates and activists, to structure and exploit the idea of climate change to do their political work for them. They have enabled climate change to emerge as the *leitmotif* of contemporary politics. These ten moves are arranged in broadly chronological order, although they interact with each other in various ways. Together, they construct an epistemic scaffold upon which rests the ideology of climatism.

Move 1: *The history of climate was reduced only to its physical history.*

Climate is a complicated idea that has meant different things to human societies in the past, and still does for diverse cultures in today's world. I have explored some aspects of this cultural history of climate in an earlier book, *Weathered: Cultures of Climate*.[5] Until the latter decades of the twentieth century, the study of climate's past would have engaged the humanities disciplines as much as it would the natural and physical sciences. It would have sought to understand how the idea of climate emerges from people's lived experiences of weather-in-place, and how notions of a changing climate are culturally and historically shaped. For example, the historian William B. Meyer tells the story of the Oneida community's quest in the middle of the nineteenth century to organize their way of life to suit the climate of New York State.[6] The Oneidans' religious beliefs undermined any idea that human attributes were determined by climate. Their view was that 'bad' weather events were much a consequence of poor planning and social organization as they were a result of a capricious agent, whether that be God or nature.

This way of thinking about the relationship between climate and people became marginalized as the twentieth century progressed and drew to a close. Instead, the natural sciences promised to explain the physical processes governing the world's weather and, consequently, promised to predict the future of climate as an aid to rational planning.

Consequence: This 'naturalization' of climate is not necessarily for the worse. But it has had one unfortunate consequence. It has meant that sight has been lost too easily of how climate and its changes have historically been blamed – at different times, in different cultures and for different reasons – for everything ranging from human physiology and war to economic success and societal collapse. What was largely erased from the scientific and social scientific study of climate was how the idea of climate has always been bound up with the human imagination. Forgotten was that this malleability of the idea of climate had enabled it historically to be enlisted in support of many different political projects – both the good and the bad, whether depictions of healthy climates or justifications for racism. The scientized idea of climate was thus stripped of cultural meaning.[7]

Move 2: *Predictive models of Earth System science began to overwhelm other approaches to the study of climatology.*

For much of the twentieth century, the scientific study of climate was rooted in statistical, regional and applied climatology. The ambitions of such inquiries were, for example, to maximize climatological knowledge for forest and agriculture management, to ascertain river flow return-periods for water resources and flood defence planning, or to delineate the nature of regional climates for purposes of economic development. Starting in the 1960s, and accelerating into the 1980s, the new technologies of computer simulation and Earth observation from space forged a new integrated, numerical and systems view of the Earth and its climate. This culminated in NASA's vision for a new science – an Earth System science – and its associated comprehensive numerical simulation models. Its goals were 'to obtain a scientific understanding of the entire Earth System on a global scale by describing how its component parts and their interactions have evolved, how they function, and how they may be expected to continue to evolve on all timescales'.[8]

Consequence: Climate became globalized in a new way, seen as a single universal system that could be simulated with – it was believed – increasing degrees of realism, and made predictable. Thus, NASA again: 'New models of the Earth System are now being developed to explore the interactions among the Earth's components and to analyse the global effects of physical, chemical and biological processes . . . these new models will also provide predictions of the effects of global change on human populations.' Global kinds of climatic knowledge – knowledge detached from specific cultural meanings – began to become dominant. This long-standing promise of prediction is alive and well today, as illustrated by the EU's 'Destination Earth' project. This project aims to develop by 2030 a highly accurate digital model of the Earth to monitor and predict with unrivalled precision the interaction between natural phenomena and human activities.[9]

Move 3: *Global temperature was adopted as the dominant index for capturing the condition of all climate–society relationships.*

We have already seen how Nordhaus pioneered the use of global temperature in the 1970s to conduct the first economic analysis of climate/energy policy. Some scientists had been thinking in terms of 'the Earth's temperature' since the nineteenth century. But they had done so in terms of the radiation physics of the world's atmosphere, not in terms of the relationship between climate, people and society. And for most of the twentieth century, scientists had struggled to derive 'global temperature' from empirical observations, as opposed to determining it through theoretical calculation. This began to change during the 1980s. Global temperature began to be employed not just as an index useful for climate physics, but as a proxy for describing the health of all climate–society relations. In 1992, in the declaration of the UNFCCC, the United Nations had left its definition of what constituted 'dangerous' climate change undefined. But four years later, in 1996, the European Union became the first political entity to adopt a climate policy target expressed solely in terms of global temperature: 'the Council of the European Union believes that global average temperatures should not exceed 2°C above pre-industrial level'. By the early years of the present century, global temperature had gained political and popular eminence, far exceeding Nordhaus's earlier use of it alongside GDP to determine rational control strategies.[10]

Consequence: From the mid-1980s, global temperature started to structure both the science and the politics of climate change, and to shape the way people thought and talked about climate change. As mentioned earlier, global temperature became a fetish. Just as it became seemingly impossible to talk about the health of economies without referring to GDP, it now became seemingly impossible to talk about climate change without referring to global temperature. And yet, as with GDP, and as we shall see in later chapters, global temperature is a seriously flawed index for capturing the full range of complex relationships between climate and human welfare and ecological integrity.

Move 4: *The 'social cost of carbon' was created to integrate climate change into economics.*

Whereas Nordhaus in the 1970s was estimating the cost of fuel-switching away from fossil-carbon energy technology, within a few years economists were asking different questions about climate change. During Ronald Reagan's two terms as US President in the 1980s, cost-benefit analysis of public policies became mandated in the United States, and increasingly in other jurisdictions. Under such regulatory regimes, economists and decision-makers needed to know how much 'damage' might be caused by emitting carbon dioxide into the atmosphere, and thereby changing the climate. The concept of the 'social cost of carbon' (SCC) was therefore developed during the 1990s. This quantity describes in monetary terms the economic damage that might result from emitting one additional ton of carbon dioxide into the atmosphere. It is a highly contentious number, influenced by many factors – ethical, political, scientific. For example, the SCC adopted by the United States' federal government under President Trump was $7 per ton; under President Biden it rose to $51.

Consequence: Guided by Earth System models and the global temperature index, the SCC connects carbon dioxide emissions to economic price signals. By offering a measure of what economists call the 'externality' of carbon dioxide emissions, the SCC has paved the way for a range of new climate-economy modelling techniques. Since the late 1990s, the SCC has also stimulated the emergence of a variety of carbon-trading schemes around the world and is one way of putting a price on carbon. By introducing the cost of carbon dioxide emissions to markets, the social cost of carbon is one important way in which the existing economic system can 'internalize' climate change.

Move 5: *How the future is imagined was reduced to the future condition of climate.*

Based on prevailing environmental conditions, particularly climatic conditions, past societies have often developed 'comforting' explanations – and hence 'self-evident' justifications – for the superiority of certain imperial races, cultures and economies. This is the ideology of environmental determinism, which historians had largely discredited by the latter decades of the twentieth century.[11] As the century drew to a close, however, a new neo-climatic determinism began to gain ground, what I call 'climate reductionism'. Climate reductionism asserts that because future climate is deemed predictable – for example, through the Earth System models mentioned in Move 2 above – the future as a whole can be predicted. Even though many other dimensions of change affecting human and ecological futures are inherently unpredictable – for example, technology, wars, pandemics, economic development, religions, cultural values – by elevating a 'predictable' climate as the primary driver of change the future can nevertheless seemingly be made knowable. The future is thus 'climatized'.

Consequence: Climate reductionism provides simple answers to complex questions about the relationship between climate, society and the future. Yet it is one way in which model-based descriptions of putative future climates assert power in political and social discourse. Such reductionism marginalizes other dimensions of change, downgrades human agency and constrains the human imagination. If climate determinism is a limited form of reasoning for explaining the past, then climate reductionism is even more inadequate with regard to foreseeing the future. There are other means of envisioning possible futures beyond relying on predictions from climate models.

Move 6: *Human adaptation to weather extremes and climatic variability was made contingent upon precise predictions of the future state of climate.*

During the 1990s and 2000s, a new way of thinking about how societies should adapt to climate emerged.

Climate reductionism (Move 5), fuelled by the promise of more powerful and more precise models (Move 2), generated the mantra of 'predict then adapt'. The thinking took hold that precise predictions of future climate were essential for making good adaptation decisions. Alice Hill at the Council on Foreign Relations in the US captures this sentiment when she claims recently that, 'without information about where and how damaging [climate] events are likely to unfold, choosing the right adaptations to invest in is little more than guesswork'.[12] This move overturned established modes of social learning about adapting to extreme weather, whether this be in traditional or modern societies. For example, Sudanese pastoralists, after many generations of reading the sky and the land in their semi-arid climate, are able to anticipate rain events and to move their livestock to locate good pastures. Or so-called Norwegian 'snowmen' use local and historical experience of anticipating impending winter weather and avalanche risk to inform the Norwegian Public Road Administration about highway clearing operations. The alternatives for anticipatory adaptation need not lie between 'accurate scientific prediction' on the one hand and 'guesswork' on the other.[13]

Consequence: Two things follow if societal adaptation to future weather extremes and climatic variability is made dependent upon scientific predictions of future climate. The first is that ever-greater precision will be demanded of climate models and their predictions. But precision is not accuracy. Models with ever greater precision offer the illusion of ever greater accuracy. Resources will be poured into the Earth System modelling enterprise that has inherent limits to its predictive accuracy at ever finer spatial and temporal scales (see Move 2 above). This explains a recent call from climate scientists for a new sustained investment of $200 million per year to develop what they call 'k-scale global models' to 'solve many of the problems standing in the way of reliable predictions of regional and local climate change'.[14] The second consequence is perhaps more worrying. Adaptive strategies and interventions will be designed with a view to 'optimizing' planning decisions and investments on the basis of inherently uncertain climate predictions. This is instead of seeking 'robust' adaptation designs which hedge against a range of different possible future climates.

Move 7: *An 'allowable carbon budget' was adopted as a proxy for securing global temperature.*

Towards the end of the 2000s, a change occurred in the way in which future emissions of greenhouse gases were imagined by scientists. Rather than focusing on different emissions pathways into the future – for example, 'high' or 'low' emissions *paths*, the focus of most previous scenario thinking – the argument gained traction that it was the total *cumulative* emissions of carbon dioxide that would largely determine future global temperature. It was thence only a short step from cumulative emissions to the notion of an 'allowable carbon emissions budget' for a given temperature target. In other words, the question could now be asked, 'How much of this emissions budget is there left to "spend" if 2°C, or 1.5°C, of warming is the desirable global temperature?'[15]

Consequence: The idea of an allowable carbon emissions budget had important consequences for policy framing and for the wider cultural imagination around climate change. By being linked to a global temperature threshold, the budget metaphor stimulated the imaginary of 'exhausting' or 'exceeding' the allowable emissions. After one's budget is fully used, there is nothing left to spend: there is no possibility of emitting any more carbon dioxide without breaching one's stated target. (This thinking had other implications, as we shall see in Move 10). As a consequence, in order to 'expand' the carbon budget to prevent its exhaustion, the idea of carbon dioxide removal became much more attractive. By taking carbon dioxide back out of the atmosphere, by whatever means, the allowable carbon budget could be expanded. By the mid-to-late 2010s, the new policy imaginary, or international norm, of 'net-zero emissions' – or the short-hand, Net-Zero – was born.[16] Net-Zero now became a proxy for global temperature, itself a proxy for human welfare (see Move 3).

Move 8: *Specific weather events started to be explained in terms of their 'naturalness' and 'humanness'.*

In the years around 2010, a new science of weather event attribution began to become institutionalized. Programmes of scientific research were established with a view to differentiating recently occurred weather events – such as specific heatwaves, droughts and rainstorms – according to their degree of 'humanness'. This science has grown in prominence such that it is now quite common to read, for example, that because of climate change Hurricane Harvey's record rainfall in Houston, Texas, in 2017 was three times more likely than that emanating from an equivalent storm in the early 1900s, or that climate change made a 2022 heatwave in Pakistan thirty times more likely. Extreme weather events are thus no longer to be regarded as natural occurrences – as 'Acts of God' – but rather as meteorological phenomena 'caused' by differing, but precisely quantified, degrees of aggregated human culpability (see the discussion about Figure 3 in Chapter 1).

Consequence: On the one hand, weather event attribution has been used to emphasize for public audiences that weather extremes are being influenced by human factors; climate change is not just something waiting to happen in the future. On the other hand, this Move introduces into the human imagination a new categorization of the world: 'bad' weather can be differentiated by precise degrees according to the relative influence of natural or anthropogenic factors. The implication is that victims of 'human-caused' weather are entitled to claims for compensation or reconstruction, potentially awarded through the newly proposed 'loss and damage fund for vulnerable countries' or through the rising numbers of climate change litigation cases. On the other hand, the victims of 'tough-luck' weather are not so entitled.[17] This Move also feeds into the narrative (exemplified above in Move 6) that robust adaptation to weather and climate extremes *requires* these types of forensic scientific analyses. Yet when adapting to flood, drought or heat it makes little difference if the consequent meteorological extreme is either 70 or 95 per cent attributable to human agency. The material force of the event remains the same.

Move 9: *The impacts of specific extreme weather events are attributed (proportionately) to climate change.*

This relatively recent move has gained momentum from the development of weather attribution science described in Move 8. But rather than simply attributing the likelihood of the meteorological extreme to human-caused climate change, the additional move is made to allocate a proportion of the resultant *damages* – whether economic costs or mortality rates – to emissions of greenhouse gases. Claims are starting to be made that, for example, 37 per cent of deaths in a specific heatwave were 'caused' by climate change, or that, in October 2019 when Typhoon Hagibis hit Japan, 40 per cent of the ensuing $10 billion worth of damage could be attributed to climate change.[18]

Consequence: These efforts to attribute damage to climate change open up the possibility of holding specific companies or countries, who may have produced or consumed specified quantities of fossil fuels, accountable for having 'caused' specific deaths or damage. This move is also motivated by the desire to underwrite scientifically the new political ambition, agreed in November 2022 at COP27 in Egypt, to establish a 'loss and damage' mechanism under the UNFCCC. But this move is only a variant of the long history of naturalizing what are primarily social disasters. It seeks to isolate the role of human agency in causing only the meteorological hazard component of any resulting damage. By emphasizing meteorology, this move implicitly ignores local or historical human agency in conditioning the economic, social, technological or political components of the disaster. As we saw in the previous chapter, it is these latter components which determine social vulnerability to a meteorological hazard, and hence the extent and distribution of resulting damage, even to a hazard that itself might be partly attributable to human causes. This move therefore greatly oversimplifies the challenges of damage attribution.[19] For example, what proportion of the devastation caused by the Syrian civil war should be attributed to President Assad's policies as opposed to the precursor drought, only a fraction of which might itself be attributable to greenhouse gas emissions?

Move 10: *The window for political action on climate change became framed according to ever-shortening timelines.*

The early 2000s established a pattern of framing climate change communications in which policy actions for addressing climate change, notably a reduction in greenhouse gas emissions, had to be achieved by a specific date. If not achieved by that date, then it was deemed 'too late'. Thus 'one hundred months' (initiated in 2006), 'twelve more years' (in 2018), 'the next eighteen months' (in 2021), 'time is short' (in 2022) have all been offered as deadlines with (presumably) the intention of accelerating the enactment of climate policy. Change must be enforced within these windows of time, it is claimed, after which it becomes too late for effective political action. Often, these deadlines have been accompanied by literal or metaphorical clocks ticking down to the decisive date. Such deadlines for political action have nearly always traced their legitimacy back to particular interpretations of climate science, often benchmarked against statements derived from IPCC reports. And they gain rhetorical power from the idea of an allowable carbon emissions budget (Move 7 above).

Consequence: The recurring trope of 'time is short' feeds a discourse of scarcity in the public politics of climate change. Time is always short, action is always urgent, the time to act is always now. This framing of time scarcity has significant political and psychological implications. The climate future is understood only in terms of a threshold, a 'point of no return', after which political action becomes too late. If time is short, then any action will do, so long as it reduces emissions. But 'doing whatever it takes', without wider considerations of the consequences, has dangerous ramifications (see Chapter 5). Not surprisingly, making time scarce leads to short-term thinking. From a psychological perspective, the effect of such deadlines is to inhibit cognitive capacity to imagine a future beyond the deadline. It also induces emotions of panic, fear and disengagement among different publics as 'the end' is imagined to be approaching.[20]

Summary

These ten moves summarize the evolution of the scientific and social scientific study of climatic change over the past half century. They also gesture to an even longer, but often forgotten, history of studying climate and its changes. Each move involved different combinations of epistemic and political motives driving it forward but, taken together, I argue that these ten moves provide the scientific scaffold upon which the ideology of climatism rests. Adopting GDP as the primary measure for describing the health of the economy makes it appear entirely natural and necessary to adopt policies which maximize GDP. In similar vein, uncritically adopting all the developments and framings of knowledge described above, and fetishizing global temperature or Net-Zero targets, makes the ideology of climatism appear scientifically grounded, epistemically justified and hence inevitable. It explains why the adoption of the slogan 'Listen to the scientists' has become so widespread in the public politics of climate change.

RETORT

Why Climate Science is Not Wrong, *But* . . .

My argument in this chapter should not be interpreted in any way as mounting a challenge to the solidity of the scientific evidence that human influences on the physical climate are now substantial and still growing. The scientific efforts that have made this knowledge credible and robust have a 100-year history behind them. In the earlier part of my career, I contributed to these efforts. For example, I spent many years compiling and analysing historical climate data from around the world and comparing the resulting trends with those emerging from climate models. For these efforts I received a personalized certificate recognizing my contribution to the joint award of the 2007 Nobel Peace Prize to the IPCC.

However, there are different ways in which scientific and social scientific knowledge about climate change and society can be crafted and assembled. There are different ways in which the story of climate change can be framed. And different framings of knowledge condition how different political actions become, first, imagined and, then, deemed possible and effective. If effective adaptation *is* dependent upon accurate climate prediction, then maybe it is best to invest in improved prediction. If there really *are* only seven more years before the world slides over a tipping point, then best to declare a climate emergency. My point in this chapter is, first, to recognize these different framings of knowledge so as, second, to be in a position to ask whether these framings are justified, and whether their effects on political action are desirable or effective. How knowledge of the world comes into being is never neutral with respect to how that knowledge is used to subsequently change the world.

The United Nation's IPCC has been an important part of this story. The IPCC maintains its public stance of offering decision-makers only *policy-neutral* knowledge and advice; it claims never to be *policy prescriptive*. In practice, however, by reifying and promoting influential scientific concepts such as those examined in this chapter, the IPCC has become a powerful agent that implicitly guides policy debates in certain directions. Since 1988, the IPCC has done more than just adopt or reflect a particular scientific way of thinking about climate change, society and the future. It has actively brought this way of thinking into being.[21] The epistemic authority acquired by the IPCC has shaped political thinking and public discourse in profound ways. It has, even if unintentionally, made possible the sequential steps described in Chapter 1: from studying climate and its changes, to the processes of climatization, to the subsequent emergence of climatism.

Having explained the epistemic and discursive conditions necessary for the emergence of climatism over the past forty years, there is now a further question to be asked. Given the dominant role afforded to science in the climatist narrative – be it the scientization of the study of climate, the call for ever more precise modelling, or statistical analyses to attribute extreme meteorological events to human-induced climate change – to what extent has the scientific and academic world assimilated, even if subliminally, this framing of climate change, this way of making knowledge about the world? Or, to pose the question more bluntly: to what extent can the climate sciences and social sciences be said to be climatist? Answering this question is the task of the next chapter.

3

Are the Sciences Climatist?

The Noble Lie and Other Misdemeanours

In December 2017, and exactly two years to the day after the signing of the Paris Agreement on Climate Change, the UN's Environment Programme convened another meeting in Paris, a 'One Planet Summit'. The ambition of this gathering was to bring together local, regional and national leaders, and those working in public and private finance, to discuss how they could 'support and accelerate global efforts to fight climate change'. At the Summit, eight central banks and financial oversight bodies established a Network of Central Banks and Supervisors for Greening the Financial System (the NGFS). This new voluntary network sought to enhance the role of the financial system in managing climate risks and mobilizing capital for low-carbon investments. Since 2017, the membership of the Network has grown and now stands at 116 members, including bodies such as the Abu Dhabi Financial Services Regulatory Authority, the Bank of Indonesia and the European Central Bank.

Back in Chapter 1 we met the 'disgraced' Stuart Kirk, ex-head of asset management at HSBC who resigned in July 2022 following what was perceived to be his public attack a few weeks earlier on some of the tenets of what I call climatism. One of Kirk's complaints was about some of the unrealistic assumptions used by central banks to calculate future financial risk exposure to climate change and how these assessments get used in environmental and social corporate governance (ESG). Amongst the more egregious examples of questionable assumptions are those embedded in the climate scenarios used by the NGFS to guide their climate risk stress-testing.

One of the functions the NGFS has taken upon itself in the last five years is to advocate the stress-testing of financial institutions to assess their financial resilience 'to hypothetical, extreme, yet plausible [climate] scenarios'. What, however, is meant by an 'extreme, yet plausible' scenario? In their scenario work from 2021, the NGFS recommended financial institutions use three different scenarios in their stress-testing. One of these scenarios was called 'Orderly', in which there is early and ambitious action to move the world to Net-Zero, and one was called 'Disorderly', in which this action is late and disruptive. The third scenario is called 'Hothouse World', in which there is very little climate policy development at all. These different scenarios carry different implications for the risk exposure of financial institutions to climate change, and hence for projected rates of return on investments.

In the 'Hothouse World' scenario, carbon dioxide emissions grow to reach 81 gigatonnes of carbon (GtC) annually by 2100. This compares with 33GtC in 2020 and is more than three times the estimated annual 2100 emissions (c. 25GtC) assuming no advance on the climate policies and technologies of 2019, a very conservative assumption. So, yes, the 'Hothouse World' scenario being used to evaluate the financial resilience of institutions to future climate risk is certainly 'extreme'.

But can it really be said to be 'plausible'? Almost certainly not. As policy analyst Roger Pielke Jr pointed out, commenting on the Kirk resignation: 'The misuse of extreme scenarios is endemic in the ESG community and beyond when discussing climate and transition risks. It is right to highlight these issues, because the misuse of scenarios is itself a major risk to global finance.'[1]

The systematic use of implausible emissions scenarios – and hence implausible climate scenarios – is not unique to the financial sector, as we will see below. This is a danger that the ideology of climatism is particularly susceptible to.

The use and misuse of climate scenarios

Climate scenarios of the future have become central features of the 'evidence base' which is offered by climate science for ascertaining how serious future climate change might be. Climate scenarios rely on a combination of trends, assumptions, judgements and models and are designed to offer plausible descriptions of how the future climate may evolve. They are used in turn by a range of scientists and social scientists to estimate possible social, economic and ecological impacts that might arise at different scales from future climate change. For example, studies might evaluate future exposure to water scarcity and heat stress, vector-borne disease, or coastal and fluvial flooding, or they might project the impacts of climate change on agriculture and the economy. These 'impact analyses' are influential in informing the development of climate policies for both mitigating climate change and adapting to it. Climate scenarios and their various applications profoundly shape the content of reports from the IPCC. They have become ever more widely used by policymakers and policy advocates, think-tanks, humanitarian organizations, banks and other financial institutions, private sector companies, and many more entities besides. And climate scenarios substantially influence the way in which climate change is talked about in public.

There is a problem, however, as we have seen in the case of the NGFS. Many of the scenarios used for informing these policy analyses and forming public opinions over the past decade have overestimated the likely magnitude of future climate change. This bias in climate scenarios has developed for two main reasons: one related to the assumptions made about how greenhouse gas emissions, especially carbon dioxide emissions, will evolve in the future; the other related to deficiencies in the climate models which convert these emissions scenarios into climate predictions. Let us explore these in turn.

For over ten years now, researchers – including the IPCC – have been using a set of emissions scenarios labelled Representative Concentration Pathways (RCPs). (The word 'Concentration' refers to the concentration of greenhouse gases in the atmosphere.) There are four main RCPs. They range from a future world in which global warming likely remains well below 2°C (RCP2.6), to a future world in which fossil-fuel burning, especially coal, continues largely unabated and in which warming likely reaches 4° or 5°C by the end of the century (RCP8.5).[2] These different RCPs were not developed as 'business as usual' scenarios – they were simply projections of future greenhouse gas emissions based on assumptions that at the time, around 2010, were deemed to be plausible. But RCP8.5 has come to be used by many analysts as the default baseline case: 'this is how the future will turn out if nothing is done'. Policy analysts Roger Pielke and Justin Ritchie showed that between January 2020 and June 2021 more than 7,000 scientific papers were published using RCP8.5, substantially more than for any of the other, less extreme, RCP scenarios.[3]

Yet the '8.5 scenario' is now the least likely of the RCP scenarios to arise. The assumptions underlying RCP8.5 – made in the late 2000s – have become increasingly far-fetched as the last decade has unfolded. RCP8.5 assumes that around 8,000 gigatonnes of additional carbon will be emitted to the atmosphere this century. This is an amount around three times greater than the RCP scenarios that most closely match current trends and policies, which are somewhere between 2,000 and 4,000 gigatonnes. More specifically, RCP8.5 assumes a fivefold increase in coal use between 2011 and the end of this century, even though coal burning, globally, peaked around 2015. With the falling costs of renewable energies, this trend of reduced coal use is unlikely to reverse, regional wars and energy security notwithstanding.[4]

The second reason for bias in some climate scenarios concerns the climate models that are used to convert these emissions scenarios into predictions of future climate. Several of the models used by the IPCC in its 2021 Sixth Assessment Report run 'too hot'; that is, they are too sensitive to future increases in greenhouse gas concentrations. This problem was recognized and reported by the IPCC in its 2021 Report. But it was much less widely noted by other scientists, social scientists or lobbyists who use these models in their impact studies or policy advocacy. As remarked recently by a group of leading climate modellers: 'We are concerned that ... much of the scientific literature is at risk of reporting projections that are inconsistent with the approach taken by the IPCC, and that are overly influenced by the hot models.' Science journalist Paul Voosen put it more bluntly: 'Many of the climate models [used by the IPCC] have a glaring problem: they predict a future that gets too hot too fast.'[5]

Not all climate scenarios combine hot models with the RCP8.5 emissions scenario. But when they do, they result in a systematic bias in the reported depiction of the rate of future climate change and in the severity of future climate impacts. Findings which suggest the future impacts of climate change might be 'worse than previously thought' can often be traced back either to the use of the RCP8.5 emissions scenario and/or to the use of hot models.[6] This might not matter much if such scientific analysis was presented in neutral terms, for example if it were used for identifying arbitrary sensitivities, testing hypotheses or furthering scientific understanding. But such analyses become misleading when their results are taken literally, when they are believed by publics and policymakers to be describing a real future ahead. These exaggerated studies become dangerous when they are used to design public adaptation and mitigation policies, for example when they are used for financial stress-testing by central banks or for estimating the social cost of carbon for economic analyses.

This example shows how the climate sciences can inadvertently sustain a climatist ideology, one in which risks are exaggerated and in which climate is elevated to be a more dominant factor shaping the future than is warranted. A few years ago, social scientist Keynyn Brysse and colleagues argued that climate scientists are biased towards issuing 'cautious estimates' of future climate impacts, erring 'on the side of least drama' as they called it.[7] My argument here is that climate scientists may very easily err of the side of greatest drama, almost without realizing it.

Science and ideology

How has climate science ended up generating predictions of future climate impacts that are 'biased high' and hence potentially misleading for policy? Might this be because climate science finds itself, willingly or not, enrolled on a particular ideological mission or, at the very least, susceptible to being influenced by the ideology of climatism? In the previous chapter, I showed how the emergence of climatism rests on ten 'moves' made in the scientific and social scientific study of climate over the past half century. In this chapter, I draw attention to the possibility of a reflexive relationship existing between the scientific knowledge being created about climate change and the ideology of climatism. The possibility I suggest – and, if so, the danger – is that science and ideology have become mutually self-reinforcing.

This would not be the first time that science and ideology find themselves in a close and interdependent relationship. An egregious example is that of eugenics, literally the study of 'good birth'. During the early decades of the twentieth century, eugenics captured the imagination of many western scientists, medics and politicians alike. Eugenicists argued that scientific techniques for breeding healthier animals should be applied to human beings. The palaeontologist and co-founder of the American Eugenics Society, Henry Osborn, pronounced thus in 1921: 'As science has enlightened government in the prevention and spread of disease, it must also enlighten government in the prevention of the spread and multiplication of worthless members of society.'[8] Mainstream scientists turned their work to supporting this ideology, holding conferences, publishing papers, securing research funding and advocating for sterilization laws.

A related example is the ideology of racism. Racism is the settled belief that racialized groups of people are physiologically endowed differently and that individuals belonging to such named groups can be ranked, judged and treated accordingly. As has been widely shown, biological science has often given support to this ideology, claiming that people can be studied and understood scientifically through the idea of 'essentialized races'. Today, this is rightly called out as racism, and it demands careful scrutiny of science's past. At the very least, being aware of such a close relationship between science and ideology, between biology and racism, raises questions about how the idea of race should and should not be used today in biomedical and genetics research.[9]

In a similar vein, this chapter explores the extent to which climate science and climate scientists have become influenced – whether inadvertently or not – by the ideology of climatism. To draw the parallel with biology and racism: might studying future societies, regions and cultures through the idea of 'essentialized climates' lend support to the ideology of climatism? By 'essentialized climates' I mean the scientistic notion that climate can be separated from the social, economic, cultural and political conditions that shape the multitude of ways in which climatic phenomena interact with the human and non-human world, and hence how these atmospheric phenomena gain significance and meaning. Just as the idea of 'race' is biocultural – race cannot be reduced to either biology *or* culture – so the idea of 'climate' should be understood as socio-physical.

The meaning of climate for societies therefore lies not simply in a series of atmospheric impulses that can be modelled by mathematical physics. Its significance rests in the interplay *between* the elemental forces of the atmosphere *and* the materiality of the ecosystems, embodied lives and infrastructures upon which these forces work *and* the human imagination that brings sense and meaning to the whole experience. Knowing the physical climate of a place no more determines the social or economic attributes or cultural worth of that place than knowing the genetic profile of an individual determines the physical or intellectual competences or moral worth of that individual. It may be possible to generate a series of numbers that describes the future state of the atmosphere. But it would be deeply misleading to plug those numbers into an algorithm that converts them into, for example, the future frequency of racist tweets in circulation, the prevalence of future violent conflicts or the economic productivity of a region. This would be just as misleading as using an individual's genetic profile to generate a narrative script explaining that individual's life history and predicting their future. This essentialism of climate has occurred as the result of Moves 1 and 2, which I introduced in the previous chapter. The idea of climate became detached from understanding the human world; and understanding climate's past and future was reduced to understanding only its *physical* past and future.

Do not misread me here. I am not equating climate scientists with eugenicists nor climatism with racism. There are many differences between these ideologies, practically, politically and morally. And therefore many differences in how the sciences underlying them may be deemed either racist or climatist. But I believe the comparison is instructive. It is at least worthwhile to explore parallels in how the structuring of scientific and social scientific investigations – whether in climate science or biological science – might lend credence to certain ideological claims about physical causation, whether these claims be that the fate of human society rests in its future physical climate, or that the fate of individuals rests in their biologically distinct genes.

I start then with the proposition that climate science and social science are, at the very least, vulnerable to reflecting, if not actively reinforcing, the ideology of climatism. By this I mean that the sciences and social sciences that underpin climate research are subject to both covert and overt pressures to lend their epistemic weight to climatism. This is a minimalist claim. These pressures might arise either internally or externally to the institutions and practices of science. In what follows, I offer a number of examples of how climate science bends, buckles and, maybe occasionally, breaks under ideological pressure. These examples concern (i) asymmetries in the evaluation and communication of climate impacts; (ii) the communication of uncertainties and the silences of climate science; and (iii) the self-interest of the climate sciences.

Asymmetrical evaluation of climate impacts

I earlier highlighted the biases in climate impact analyses that might result by combining implausible emissions scenarios with hot climate models. But biases can still arise even if analysts select more plausible emissions futures and use climate models that are better constructed to reproduce observed warming to date rather than running hot. Climate researchers might find themselves lending support to climatism either through the practice of climate reductionism or through evaluating climate impacts asymmetrically.

Climate reductionism isolates the impacts that changes in climate might have on the future world and largely ignores the many other changes in society that will also be occurring, whether in technology, geopolitics, economics or cultural values. This was presented as Move 5 in the previous chapter. The result is that climate is given an outsized role in shaping the future. This feeds the ideology of climatism. For example, take the future of malnutrition, malaria, diarrhoea and heat stress. In 2021, the World Health Organization calculated that by 2050 'climate change will cause 250,000 additional deaths globally per year' from these various diseases.[10] A changing climate undoubtedly alters the patterns of disease transmission and will impact upon people's diet and hence their susceptibility to nutritional diseases. But climate is far from being the only factor, or even the dominant factor, in determining the future burden of communicable and nutritional disease. Economic trends, wealth distribution, lifestyle choices, rural-urban mobility, educational attainment, public health infrastructure and educational programmes, and many other factors besides, will have a much bigger impact on these future disease burdens. Claiming a specific number of 250,000 such deaths as being attributable to 'climate change' is little more than whistling in the dark. It implies that if future changes in climate can be arrested over the next years then 250,000 lives will be saved. This is misleading.

This is offered as just one example of climate reductionism, but it illustrates how the sciences can easily sustain the ideology of climatism. A second, but related, problem concerns the inventories, evaluation and communication of climate change impacts. Whilst there is no doubt that a warming climate will induce a wide range of serious risks and challenges to human welfare and ecological integrity, not all the impacts of climate change will be uniformly negative for all regions or sectors. Even after allowing for the problem of climate reductionism, the systematic evaluation and public communication of future impacts that appear in scientific assessment reports do not always reflect this reality.

This particular bias was highlighted in 2010. Following the earlier publication in 2007 of the IPCC's Fourth Assessment Report, a small number of errors or inconsistencies were identified in the element of the Working Group 2 Report that focused on climate impacts, adaptation and vulnerability. One of the most egregious factual errors concerned the Netherlands, with the IPCC stating that 55 per cent of the country was below sea level. The correct figure is only 26 per cent. The Dutch government were intensely annoyed and commissioned the Dutch Environmental Protection Agency (PBL) to conduct a full line-by-line audit of the Report to root out other errors. The audit identified a relatively small number of rather vague or unsubstantiated statements, but the basic conclusion of the Report was upheld, namely that many future impacts of climate change were serious.[11]

However, the audit also found that in the Summary for Policymakers (SPM) only the negative impacts of climate change were highlighted. There was no mention of any benefits of climate change. The IPCC authors defended this selection as being consistent with taking a 'risk-oriented' approach to climate change assessment. They claimed that this was what the world's governments had asked them to undertake. On the other hand, the PBL audit criticized the IPCC for not being explicit about the approach it had adopted. An uninformed reader of the SPM – often the only part that is read by policymakers or by interested publics – would have concluded that there were no benefits of climate change anywhere in the world. PBL's recommendation to the IPCC was that future reports should be explicit about the choice of assessment methodology and about the basis upon which impacts were selected to be highlighted. PBL's preferred solution was that future SPMs for Working Group 2 be issued in two sections. One section should describe the full range of projected climate impacts, including uncertainties, influential non-climate factors and positive impacts. The second section should then describe the most important negative impacts of climate change, following the worst-case risk-based approach. This greater transparency would enable policymakers to gain a more symmetrical view of future climate impacts. This recommendation has not been followed in either of the two subsequent IPCC reports in 2014 and 2022.

A more recent example of selective reporting of climate impacts comes from the BBC's Bitesize website. Bitesize is the BBC's website for school-age children, and one of its educational pages for fourteen- and fifteen-year-olds dealt with climate change. Although pointing out many of the risks and negative impacts of future warming, the page also reported that warmer temperatures 'could lead to healthier outdoor lifestyles' and that a warming of high-latitude climates in the Arctic region could mean easier access to oil in Alaska and Siberia. There were some other potential benefits listed, including the future extension of agricultural production in Siberia and new shipping routes through the Arctic created by melting ice. These are all uncontroversial likely consequences of climate warming. However, in response to vigorous and sustained pressure from climate scientists and advocacy groups in 2021, the BBC removed mention of such 'benefits' of climate change from these educational pages. School children were only to learn of the negative impacts of climate change.[12]

Communicating scientific uncertainties

As we saw in the previous chapter, the ideology of climatism rests heavily on a scientistic scaffold. In other words, climatism draws on the claims of the climate sciences and scientists for its epistemic and rhetorical credibility. This means that it is vulnerable to a very particular line of criticism. If the claims of climate science, or even just some of the claims, can be shown to be untrue or exaggerated then doubt can be thrown on the whole edifice of climatism. Conversely, it also opens climate science to the possibility of being distorted by political pressure. Climate scientists may find themselves defending the authority and 'truth' of climate science *at all costs*, for fear of revealing to critics, or to the public at large, awkward uncertainties and unresolved ambiguities about the underpinning science.

Yet uncertainties in knowledge – the ambiguities of evidence, and provisional and contested findings – are ubiquitous in the sciences. Indeed, being explicit and open about such features of science is one of the characteristics that distinguishes science from dogma. As we saw earlier, these features are especially present when scientists are called upon to make predictions about the future. Climate scientists may therefore find themselves caught between a rock and a hard place. The late Stephen Schneider, a prominent American climate scientist, famously captured this predicament using the metaphor of the 'double ethical bind':

> *On the one hand, as scientists we are ethically bound to the scientific method, in effect promising to tell the truth, the whole truth, and nothing but – which means that we must include all the doubts, the caveats, the ifs, ands, and buts. On the other hand . . . to reduce the risk of potentially dis-*

> *astrous climatic change . . . entails getting loads of media coverage. So, we have to offer up scary scenarios, make simplified, dramatic statements, and make little mention of any doubts we might have . . . Each of us has to decide what the right balance is between being effective and being honest. I hope that means being both.*[13]

Schneider's double ethical bind has led some philosophers to advocate that scientists embrace the equivalent of Plato's 'noble lie', the knowing propagation of falsehood or deception by an elite in order to maintain social harmony or to advance an agenda. In other words, the cause justifies the lie. For example, the veteran philosopher of science Philip Kitcher advances such an argument specifically with respect to climate science. In his 2011 book, *Science in a Democratic Society*, Kitcher puts forward this scenario: An atmospheric scientist makes a discovery that seems to challenge a particular model of sea-level increase due to global warming. She expects her discovery will be refined through further research and that, in the end, it will not refute the mainstream view. In the meantime, she wants to avoid giving ammunition to climate sceptics, so she postpones publication. But an ambitious postdoc surreptitiously informs the media about her discovery. The media accuse the scientist of a cover-up and report that key evidence for anthropogenic climate change has been refuted.[14]

Kitcher concludes that the atmospheric scientist *was* justified in withholding her discovery from the public. In other words, being 'economical with the truth' is fine. For Kitcher, the hypothetical scientist 'wisely foresaw the danger that [her discovery] would be deployed in misleading ways and attempted to do her bit for the promotion of public freedom'.

This is a dangerous position to espouse. It justifies the suppression of scientific findings for fear that, at the least, they would reveal to the public the provisional and uncertain nature of the scientific enterprise. At worst, such findings might undermine the 'good cause' that a scientist may believe their work is seeking to sustain. Either way, such a justification for the noble lie subverts professional scientific norms and practices. It runs the risk of 'bringing science into disrepute' for the sake of protecting what may be believed to be a higher ethical objective. In this scenario climate science reveals itself to be climatist: it defends the ideology of climatism rather than upholds the ideals of science.

A potent real-world example of such inappropriate defensiveness occurred late in 2009. The Chair of the IPCC, Rajendra Pachauri, dismissed a challenge from the Indian government claiming that a statement in the IPCC's Fourth Assessment Report about the melting rate of Himalayan glaciers was in error by an order of magnitude. Rather than investigating the claim, Pachauri dismissed it on live Indian TV as being 'arrogant' and representing 'voodoo science'. By using the term 'voodoo science', Pachauri sought to demarcate what he wanted to be legitimate scientific knowledge – wrongly as it later turned out – from mere belief, superstition or ideology. He falsely accused the Indian government's challenge of being politically driven and scientifically disreputable, which, ironically, turned out to be the case with the IPCC's handling of the complaint.[15]

Another example of the suppression of scientific uncertainty for the sake of a 'good cause' was the Climategate controversy which also erupted during the last few weeks of 2009. Following the unauthorized acquisition and publication of thousands of emails stored on a computer server at the University of East Anglia, the scientists involved in these email exchanges over several years were accused of professional and ethical misdemeanours: of suppressing data which were 'inconvenient' for the standard theory of climate change; of cherry-picking statistical results that were favourable to 'the cause';[16] and of perverting the process of scientific peer review. Although Climategate did not refute the basic conclusions of mainstream climate science, it dramatically revealed the pressures that climate scientists felt under to defend 'the cause' ahead of defending the integrity of science. Motivated reasoning and groupthink can infect climate science just as much as it affects other groups in society.[17]

Climate science struggles with the public communication of scientific uncertainty. Another example concerns the trope of 'only *n* years left before it's too late'. The most egregious example of this occurred in 2018 following the publication of the IPCC's *Special Report on 1.5°C of Global Warming*. A subsequent article in *The Guardian* newspaper interpreted this Report as warning that there were only '12 years [left] to limit climate change catastrophe'.[18] This claim was rapidly taken up around the world by new protest movements, such as Extinction Rebellion, but also by establishment organizations such as the World Economic Forum and the UN Secretary-General's office. But the '12-years to catastrophe' finding was nowhere to be found in the Report and, in any case, such apocalyptic pronouncements do not reflect the findings of empirical science. Rather, they are explicitly normative, and rhetorical, claims. A small handful of climate scientists challenged this deadlineism. Yet the IPCC remained silent in the face of such wilful and widespread misrepresentation of its own Report, tacitly lending its authority to the claims. As the most public voice of climate science, the IPCC implied that climate science as a whole was comfortable with the noble lie. For the sake of 'the cause', it was thought that such a social amplification of risk, masquerading as science, might not be such a bad thing.[19]

Self-interested science

There is one further area worth highlighting where the climate sciences might be susceptible to the claim of being climatist. It concerns the self-interestedness of science and scientists. Such an accusation has often been made against science in general and against specific sub-fields of science in particular. Rather than being motivated by a disinterested 'search for truth', the scientific enterprise has at times been criticized for seeking to further, or being perceived to be furthering, its *own* interests. This might include securing research funds, protecting science's autonomy or furthering individual scientists' prestige and their access to power.

The sociologist of science Brian Salter draws attention to one facet of this problem, namely over-claiming. In the fields of genomics and bioinformatics Salter observes that, 'in order to improve their fields' chances of success in the research funding market, scientists are accustomed to making promises where the delivery or timescale are problematic'. Salter goes on to explain that when promises or raised expectations of scientific advances or breakthroughs are not met, scientific domains face a common political problem, namely 'maintaining the legitimacy of the field' in the eyes of politicians and the public.[20]

Within the climate sciences the most obvious example of this pattern of over-claiming concerns the promises of climate prediction. In the previous chapter, I drew attention to this feature of climate science in Moves 2 and 5. The temptation for climate modellers to promise that another (large) injection of public funds will lead to greater accuracy in their predictions of future climate is ever-present. Greater precision, maybe. But improved accuracy is quite different from greater precision. It is often more useful to offer imprecise predictions that are accurate, than precise predictions that can be totally wrong. 'Better' climate models may well yield more knowledge about physical processes in the climate system. But more knowledge may simply lead to greater uncertainty. The latest bid for a huge injection of public funds into climate modelling concerns the development of what is being called 'k-scale climate models', where the 'k' represents a one-kilometre grid of the Earth. This represents an astonishing 2,500-fold increase in model grid resolution – and it comes at an astonishing price. As one leading climate modeller, Gavin Schmidt, put it, 'I worry that these new proposed efforts will be focused more on exercising flashy new hardware than on providing insight and usable datasets. I worry that implicit claims that climate model prediction will be as improved by higher resolution as weather forecasts have been will backfire.'[21]

This line of self-interested science has at times been developed by those wishing to discredit the entire scientific basis for the human influence on climate. Jeremiah Bohr has drawn attention to this contrarian manoeuvre by using the phrase 'the climatism cartel' to depict this characterization of climate science. In Bohr's description, contrarians accuse climate scientists of operating a cartel, a closed shop of like-minded parties who recognize in the idea of climate change a means of furthering their own mutual self-interest.[22] I am certainly not suggesting that this is necessarily the case with climate science. But I do believe that there exist pressures, both within climate science and external to it, that, at the least, can give the appearance of climate scientists being self-serving. If the climate sciences are to resist the slide into being climatist, then an openness to public challenge, scrutiny and audit is essential.

RETORT

The Integrity of Climate Science

Reading this chapter, some might accuse me of claiming that climate science is biased, misleading, untrustworthy and essentially worthless. This is not what I am saying. Rather, I am drawing attention to the following. Scientific research is always conducted within specific social and political contexts and it cannot escape the influence of interests and values, both those of the scientists themselves and those of the societies in which science is funded and is undertaken. This is something that historians, sociologists and philosophers of science have repeatedly and convincingly shown. Rather than diminish the public value of scientific knowledge, such recognition in fact frees science to take its appropriate place in the public sphere, and therefore in decision-making, alongside other forms of circulating knowledge, cultural values and political goals. Being alert to such influences can better separate the scientific chaff from the scientific grain. We can attend more carefully to those norms, values and practices that make science valuable and distinct from other human modes of knowing.[23]

For example, it helps us to see that climate science is not immune to group-think or to the allure of the noble lie, and that the best antidote to such practices is a relentless questioning of motives, assumptions and potential blind-spots. It helps us to see that climate science is likely the best way of determining the physical processes through which humans are influencing the climate, but that scientific knowledge is far from sufficient to secure climate justice, or indeed any other contested political goal. It helps us to see that communicating openly and honestly the uncertainties and ambiguities of climate science is a precondition for scientific knowledge to gain public trust and political value. Protecting the integrity of science in this way, and learning the lessons from science's past relationship with eugenics and racism, will inoculate the climate sciences against the danger of being co-opted by the ideology of climatism. If science stops being self-critical, even the science of a noble cause such as alerting the world to the risks of climate change, it stops being science. And if climate science stops being self-critical and honest with itself and its multiple publics, it starts being climatist.

Summary

This chapter has drawn attention to the danger of the climate sciences and social sciences falling into a climatist stance. By this I mean science succumbing to either internal or external pressures to lend tacit or explicit support to the ideology of climatism. These pressures are exerted by the media, by campaigners, by the salience of international political commitments and, as we saw in the case of ESG and the European Central Bank, by institutionalizing the requirement to conduct climate risk assessments across all public and private institutions and companies. Much is being asked of climate science. Perhaps too much.

I have suggested that, rather than 'erring on the side of least drama', climate science and social science have at times cranked-up the alarm through the over-promotion of hot scenarios, through the unreflexive adoption of climate reductionist assumptions, and through asymmetric treatment of climate change impacts. At other times climate scientists have failed to warn adequately when the alarm is being rung too loudly. And because the knowledge they produce is so necessary to sustain the ideology of climatism, climate scientists and social scientists sometimes find it necessary to justify the myth of the noble lie. That is, for fear of showing any weakness to their critics, some scientists have at times felt the need to defend their knowledge at all costs, even if this means suppressing or eliding uncertainties in their science.

In making this argument I have alerted the reader to the (partly) analogous case of science and racism. There has been a long history – and a history not yet fully closed – in which scientific theories, concepts, models and experiments have perpetuated the idea that race is a biologically determined category. In the process, science has inadvertently lent credence to a range of politically and ethically dubious attitudes and practices. By analogy, I offer a warning to the climate sciences: be careful lest the theories, concepts, models and experiments of climate science lend credence to an essentialist or reductionist view of climate's relationship with people and societies. This opens the door to policies that can lead to perverse outcomes, as we will see in Chapter 5.

Having established the basic contours of climatism, and its close and potentially problematic relationship with the climate sciences, I now need to answer a different question. Why has the ideology of climatism gained such prominence in today's world? In particular, what is it about the master-narrative of climate change advanced by climatism that makes the ideology appealing to so many individuals, enables it to take root in social movements, and empowers it to start shaping public institutions?

4
Why is Climatism So Alluring?
Master-narratives and Polarizing Moralism

The film *Don't Look Up* is an apocalyptic satire released in the United States and on Netflix in December 2021. It stars Leonardo DiCaprio, Jennifer Lawrence and Meryl Streep, and during Christmas week of that year it was the most-streamed English-language film on Netflix and received the second highest viewership for a debut movie during its first weekend.

Don't Look Up tells the story of two American astronomers – Kate Dibiasky, an astronomy PhD candidate, and her professor, Dr Randall Mindy – who try to alert humanity to a comet heading for Earth. The comet is of such mass as to likely destroy human civilization on impact. Dibiasky and Mindy present their warning directly to the President of the United States and her White House staff, but are met with apathy. After some prevarication, a plan to intercept and redirect the path of the comet using nuclear weapons is aborted by the President in favour of efforts, using an unproven technology, to fragment the comet so that its rare-earth minerals can be retrieved after its dispersal into the ocean. This plan also goes awry and so the comet continues on its collision course with Earth. The film's title is drawn from the President's 'Don't look up' public media campaign, launched to counter the science-inspired social movement telling people to 'Just look up' so they can see for themselves the reality of the approaching comet. As the US government elites escape on a spaceship designed to find an Earth-like planet elsewhere, the comet strikes Earth off the coast of Chile. Extinction follows.

The film sparked enormous discussion online and gained mixed reviews from professional critics. Mick LaSalle of the *San Francisco Chronicle* praised the film: '*Don't Look Up* might be the funniest movie of 2021. It's the most depressing too, and that odd combination makes for a one-of-a-kind experience ... McKay [the director] gives you over two hours of laughs while convincing you that the world is coming to an end.' Amit Katwala of *Wired* concluded that '*Don't Look Up* nails the frustration of being a scientist', while Linda Marric of *The Jewish Chronicle* gave the film four out of five stars, writing: 'There is something genuinely endearing about a film that doesn't seem to care one bit about coming across as silly as long as its message is heard.'

And what is the message? As explained by the film's director, Adam McKay, *Don't Look Up* is an overt allegory for the supposed existential threat posed to humanity by climate change. The comet's threat to life on Earth stands in for climate change and the film's plot becomes a comic satire of government, political, celebrity and media indifference – particularly American indifference – to the risks associated with climate change. But the allegory only works by assuming a very particular reading of the risks to life posed by climate change. The film is premised on the adequacy of an asteroid impact acting as a metaphor for climate change. Climate change risk is taken to be the equivalent of an Earth-shattering meteorite hit, a one-off catastrophic event which decisively re-orders planetary life and function. As one of the characters in the film says, 'We are trying to tell you that the entire planet is about to be destroyed!'

This a powerful, widely used and alluring metaphor, but it is misleading. Climate change is not like a meteorite hit. It is not a dramatic or decisive threat with existential consequences. The challenges of climate change are a lot messier, more indeterminate and with a slower unfolding time-frame. McKay's film portrays climate change as a 'simple problem' awaiting a decisive single-shot solution, delivered by technology, if only the politicians believed the astronomic science. *If only* the President would 'listen to the scientists', McKay seems to be saying. However, by offering an asteroid strike as an analogy for climate change, *Don't Look Up* fails to recognize climate change as a wicked problem, one with diffuse causes, effects that are unevenly distributed, and with no universal solution. Alex Trembath points out that this is what makes the film troubling:

> By indulging the fantasy that a few corrupt politicians and plutocrats are the principal forces blocking action on climate change, these stories – whether on or off the screen – distort the way the public thinks about the problem. It ignores the gaps in our technological capability to decarbonize the global economy, pretending that the prosperity and comfort enabled by fossil fuels could easily and quickly be provided by low-carbon technologies.[1]

Climatism as master-narrative

Don't Look Up follows in a long line of parables warning of existential threats to the future of humanity and of the Earth upon which we live. From the ancient flood myth of Gilgamesh, inscribed in the Jewish scriptures as Noah's Flood, to Rachel Carson's *Silent Spring*, the human imagination has been drawn to narratives that signal a decisive and apocalyptic ending. They are fables that point to moral degeneracy, selfish greed or technological hubris. Climatism, too, frequently displays this apocalyptic flourish and moralizing tone. As with these older parables, it also offers a totalizing master-narrative which explains the condition of the world and its likely future.

The allure of climatism therefore is that it seems to offer a comprehensive, coherent and persuasive account of the present and future state of the world. Nothing is left out. The master-narrative of climatism appears sufficiently elastic to present climate change as an explanation for nearly everything – from the success of the Taliban in Afghanistan, to inter-ethnic violence in Syria, to the decline of insect populations in Europe, dramatic Australian wildfires, the rise of gender-based violence and the precarity of the Sundarbans. It is in this sense that climatism starts to become – to adopt Michael Freeden's terminology – a macro-, or thick, ideology. It becomes an ideology that envelops and subsumes the full range of concerns about our experience of today's world and our worries about the future. And it induces emotions of, paradoxically, both anxiety and comfort. By offering a view of the future dominated by the seemingly inexorable dislocation of climates around the world – of 'climate breakdown' – a state of anxiety is induced. Yet, on the other hand, the totalizing scope of climatism's account of the world's ills suggests that *if only* climate change could be arrested or reversed then a better world would await. Or, at least, such an outcome would stop things getting worse. 'If only the President would listen to the scientists.'

This chapter lays out four general features of the ideology of climatism that help explain its appeal and wide embrace: its totalizing scope, its gnostic tone, its apocalyptic rhetoric and its Manichean worldview.

Totalizing scope

Master-narratives offer comprehensive explanations of historical experience and/or knowledge about the future. To quote two leading academics who study them, a master-narrative 'is a global or totalizing cultural narrative schema which orders and explains knowledge and experience'.[2] Master-narratives are successful if they can pull together and organize in some larger totalizing scheme a range of smaller, more fragmentary, narratives. The world we experience is complex and frequently bewildering. And given the hyper-visibility of today's world, mediated through our smartphones, we experience ever more of its bewildering complexity. And we do so instantly. Part of the allure of ideologies is that they promise decisive explanations for what may otherwise appear perplexing, random and unconnected events. This is as true for climate events as for anything else. They also offer the promise of an ultimate meaning to such events beyond their seeming randomness. Or, if not an ultimate meaning, then at least a sufficient meaning around which one can purposefully organize one's life.

The ideology of climatism is no different. It rests on a totalizing and easy-to-understand narrative about the state of the world and it offers an alluring and meaningful agenda for political action. The advertising blurb for the 8th Potsdam Summer School, held in

Potsdam, Germany, in 2022, offers one rendering of this meta-narrative:

> Humankind is facing a huge challenge: Climate change threatens the foundations of life on our planet. To preserve these foundations, the fossil era must come to an end. This will lead to profound changes in our ways of producing goods, our means of transportation, and ultimately the way in which we live. We are at the beginning of a great transformation, which can bring benefits as well as new inequalities. It is important to shape this in a just and sustainable manner, while protecting the environment and thereby securing future life for people in all regions around the globe.[3]

In just 100 words, we see here the attraction of climatism. Climate change explains the reasons for our current predicament and outlines a grand challenge according to which future political projects can and should orient themselves. The narrative is assured, instructive and normative. While not promising success, it at least describes what success would look like.

Or take the example that Naomi Klein offered in her book *This Changes Everything: Capitalism vs the Climate*. Explaining the moment when she first took climate change seriously, we see Klein describing the totalizing character of her enlightenment:

> I started to see signs – new coalitions and fresh arguments – hinting at how, if these various connections were more widely understood, the urgency of the climate crisis could form the basis of a powerful mass movement, *one that would weave all these seemingly disparate issues into a coherent narrative* about how to protect humanity from the ravages of both a savagely unjust economic system and a destabilised climate system.[4] [emphasis added]

Theologian Lisa Stenmark helps us to understand the attraction of such master-narratives, or what she calls 'myths of the Absolute'.[5] When faced with complexity and uncertainty, or situations beyond our understanding and control, the human instinct is to reach out for something beyond ourselves and our limited experience that can make sense of it all. For many, this leads to an encounter with things spiritual, eternal and transcendent. For others, this religious urge is met through embracing powerful myths or stories which suggest that we are not just an accident of history, but part of something meaningful and purposeful. Master-narratives, such as the one offered by the ideology of climatism, meet these criteria. They draw people together into collectives, where shared identities, public judgements and political actions can be forged and executed. It is for this reason that climate social movements such as 350.org, the Sunrise Movement, Extinction Rebellion or Fridays for Future have rapidly gained adherents in recent years.

Gnostic tone

A second feature of climatism which contributes to its allure is its *gnosticism*. By this I mean that climatism gains strength and appeal from the nature of the 'special knowledge' upon which it rests. Gnosticism is an ancient idea originating in spiritual and religious traditions, in which the principal element of salvation is direct knowledge of the divine, acquired in the form of mystical or esoteric insight. Gnostic texts therefore emphasize ideas of privileged enlightenment through divine disclosure. This knowledge sets believers apart from non-believers. It might be used, for example, to make authoritative claims that the future will unfold in ways that fulfil divine purpose.

I suggest that all ideologies are gnostic in the sense that they lay claim to the possession of special knowledge. Ideologies such as Marxism or Aryanism, or the 'myths of the Absolute' such as are found in monotheism, rest their authority on historical, intuitive or metaphysical revelations about the true nature of the world and its unfolding. These master-narratives hold in common the belief that otherwise bewildering phenomena – historical, social, cultural and even political phenomena – can be explained through their respective 'special knowledges'. In the case of Marxism the source of this knowledge would be historicism, for Aryanism ancient mysticism, and for the Abrahamic faiths divine revelation.

For climatism, this 'special knowledge' emerges from scientific and social scientific claims about the past and future, as we saw in Chapter 2. This scientistic basis for climatism sets it apart from most other ideologies and grants it a distinctive authority and status in most modern societies. If climatism rests on (special) *scientific* knowledge then, as the popular understanding of science would imply, it must be true and reliable. Unlike Marxism, Aryanism or monotheism, therefore, climatism becomes non-negotiable. This is why political actors defending the interests of the fossil-fuel industry focused so much of their attention in the 1990s and 2000s on seeking to discredit climate science and scientists.[6]

This feature of climatism helps explains it allure. It also explains the frequency with which questions and debates about policies to tackle climate change are deflected to 'the scientists'. The slogans are many: 'the science is in', 'the science is clear', 'scientists tell it like it is', 'follow the science', 'trust me I'm a scientist'. Scientists are the ones with the 'special knowledge', the truthful insight into the human and planetary condition. Greta Thunberg, for example, has frequently made this the basis of her defence when questioned, as in this interview for BBC radio on 23 April 2019:

> Interviewer: 'What do you want people to do, what do you want governments to do?'
> Thunberg: 'listen to the science, listen to the scientists . . .'

Similarly, a few months later, when addressing the US House of Representatives in September 2019, she again deferred to the authority of science. Thunberg said: 'I am submitting this report as my testimony because I don't want you to listen to me, I want you to listen to the scientists. I want you to unite behind science. And then I want you to take real action.'[7]

This feature of climatism is an example of scientization, when scientific statements substitute – or at least become a short-hand – for ethical or political reasoning and argument. 'Why is limiting warming to 2°C a necessary goal? Because the scientists say so.' 'Why do we have to achieve Net-Zero emissions in the next decade? Because the IPCC says so.' What are in essence value-based judgements about different goals, courses or priorities of action appear as blunt statements of fact derived from scientific inquiry and scientific truth. This defence is an attractive discursive tactic in the free-for-all of political life.

Apocalyptic rhetoric

So, climatism is alluring because it offers a totalizing narrative which explains the human predicament and because it presents itself as rooted in scientific – and hence authoritative – claims about the past, present and future. A third feature of climatism's allure is its frequent invoking of apocalyptic rhetoric. Apocalyptic rendering of environmental hazards, changes and impacts has been one of the most frequently recurring tropes of environmentalism over the past century. The American literary critic Lawrence Buell, for example, has described 'apocalypse' as 'the single most powerful master metaphor that the contemporary environmental imagination has at its disposal'.[8]

Apocalyptic renderings of climate change are powerful, but they are also alluring. There is a long tradition in European romantic thought of the idea of the sublime in nature. This is the idea that, when confronted with nature's irresistible power, the consequent feelings of awe, terror and danger exert an irresistible pull on the human imagination. In spite of the danger, we find ourselves enraptured, captivated by the forces of nature. We are fascinated by the frisson of vulnerability, destruction and death. Something of the allure of the apocalyptic is revealed by American tornado-chasers as they track down a big one – 'The storm sky's like a chorale with thunderous organ music', says tornado-chaser Anton Scimon. Others have compared the lurid descriptions of prospective climatic disruptions – 'methane fireballs', 'hypercanes', 'climate meltdown' – with the seductions of pornography.[9] Such descriptions entice audiences with the promise of vicarious experience and a sudden strong surge of dangerous excitement.

The apocalyptic rhetoric of climatism can take both visual and verbal forms. Climate-related hazards – such as hurricanes, wildfires, floods, droughts, ice-storms – offer dramatic and powerful visual narratives that are easily linked with climate change. In a communicative culture increasingly dominated by the visual, such extreme manifestations of vicarious disaster lend themselves to instant and intuitive judgements about the agency of climate. They become the massively circulating visual memes underpinning climatism. Hard to deconstruct by the individual, they become 'the face of climate change'. The detailed forensic work that I discussed in Chapter 2, of trying to attribute specific extreme weather events and/or disasters to different causes, of necessity takes place in the background. It occurs over much slower news cycles and with far less public attention.

Such *visual* rhetoric feeds some of the more egregious examples of *verbal* rhetoric associated with climatism. Take this example from veteran environment campaigner Mayer Hillman in an interview conducted in 2018:

> 'We're doomed', says Mayer Hillman with such a beaming smile that it takes a moment for the words to sink in. 'The outcome is death, and it's the end of most life on the planet because we're so dependent on the burning of fossil fuels. There are no means of reversing the process which is melting the polar ice caps. And very few appear to be prepared to say so.'[10]

Hillman's is an extreme and despairing position. Extreme positions can be effective in gaining public attention, forging collective identities and mobilizing social concern for the promotion of political action. But such extreme apocalyptic rhetoric around climate change also has significant downsides.

Take this example of the effects of the doomism that Hillman speaks of. In a video post in October 2021, twenty-seven-year-old Californian TikTok host Charles McBryde revealed that he was a 'climate doomer'. Climate doomers, drawn in by climatism's dark apocalyptic rhetoric, believe there is little that can be done to slow or to stop climate change. McBryde admitted to 'feeling overwhelmed, anxious and depressed about global warming', but he also called out to his 150,000 followers on TikTok for help to escape his addiction to such feelings: 'Convince me that there's something out there that's worth fighting for, that in the end we can achieve victory over this, even if it's only temporary.'[11]

Just as apocalyptic rhetoric around climate change can be alluring, it can also be exclusionary. It can create narrative lock-in, leaving no space for nuance or compromise. Nor indeed does it keep open the discursive spaces for deliberation or argument which are necessary to enable the difficult political work of reconciling competing interests and goals.

Apocalyptic rhetoric is also often associated with moral dualism. It reinforces boundaries between in- and out-groups, dividing the world into friends and enemies. Which brings us to a fourth alluring feature of the ideology of climatism: its promise of moral sorting.[12]

Manichean by design

Derived from the third-century Iranian spiritual writer Mani, the doctrine of Manicheanism posits a cosmology with two competing eternal powers, one good and one evil. This worldview manifests in a belief in an ongoing cosmic struggle between these powerful divinities and leads to a dualistic understanding of good and evil. When articulated and applied to the human world through specific ideologies, this moral dualism has particular appeal to human intuition: there are two sides competing for ascendency, us and them, the sheep and the goats, the angels and the devils. A Manichean worldview offers a form of moral populism, an easily applicable and comforting framework for moral sorting.

Part of the allure of climatism is that it easily slides into a Manichean worldview. It offers two clearly identified sets of protagonists who are engaged in a struggle over the reality and meaning of climate change and its possible solutions. Climatism thrives on seeing the politics of climate change in binary terms: those who are in favour of 'climate action' and those who are against. For climatists, the enemies are nefarious fossil-fuel industrialists, right-wing conservatives and self-interested populists.

The appeal of the moral dualism offered by climatism is that it provides an easy rule-book for people looking to identify friends and enemies, to separate the climate doves from the climate hawks. This task has been greatly aided over the past fifteen years by the rise of Twitter. *The Guardian* columnist Rafael Behr points to Twitter's poisoning effects on politics in general, but his description of its polarizing effects and its amplification of Manichean dualism applies equally to climate change:

> [Twitter] is a vast polarising machine – a centrifuge that separates politics into the most extreme interactions of any position. When the ideal conception of politics might be rival teams, advancing competing policy prescriptions based on some common set of facts, Twitter turns us into quasi-religious cults, looking at the world in terms of righteous believers and despicable blasphemers . . . Twitter once appeared to be a way of making politicians sound more human. Instead, by making every comment vulnerable to wilful misrepresentation, penalising loose talk and promoting diatribe over dialogue, it turns ordinary users into caricatures of the worst kind of politician.[13]

A good example of how this plays out within climatism comes from the popular writings of the American climate scientist, Michael Mann. In his most recent book *The New Climate War: The Fight to Take Back Our Planet*, Mann views the politics of climate change through a Manichean lens.[14] The source of all opposition to the 'correct' view of what should be done about climate change is traced back to an evil empire, orchestrated by the fossil-fuel industry. This industry represents, as Mann calls it, 'the eye of Sauron', the omnipotent dark power in Tolkien's *Lord of the Rings*. The need of climatism to sustain a bipolar world leads people like Mann into an ever-widening policing exercise. The simple moniker of 'climate denier' is no longer sufficient to hold his Manichean worldview intact. In *The New Climate War*, Mann therefore identifies eight alliterative groups of emerging enemies who now muster together under the battle flag of climate *inactivism*: dissemblers, deceivers, downplayers, dividers, deflectors, doomers, delayers, distractors. For Mann, the circle of enemies has grown, mutated and, perhaps most sinister of all, now infiltrated 'the climate movement' itself. The ideology of climatism has been breached from within.

There is no doubting the need for an accelerating transition away from fossil fuels. And there is also no doubt that vested political interests have obstructed this transition.[15] But Mann is so conditioned by a Manichean worldview that wherever he looks in the public, scientific and political debates around climate change he sees the shadows of the fossil-fuel lobby. People with whom Michael Mann disagrees – a long list that now includes the likes of Michael Moore, Bill Gates and Naomi Klein – become enemies, agents of the dark forces of inactivism, or contrarians, or 'soft denialists', or deflectors or apologists or defeatists. Mann might be an extreme voice promoting climatism, but he has 215,000 followers on Twitter. His playbook here is reminiscent of 1950s McCarthyism or the Soviet interior ministry's ideological purification of the Communist International in the 1930s during the Spanish Civil War.

RETORT

What Sort of Story *is* Climate Change?

Some might take my argument here to be implying that because climatism offers an alluring master-narrative it must be wrong. Actually, in one sense I am saying just the opposite. If we understand that the credibility of such narratives lies in their ability to appeal to human needs, instincts and desires, then the master-narrative offered by climatism makes it powerful and therefore, in some sense, 'true'. Many centuries ago, Plato in *The Republic* recognized the performative power of stories and, for that reason, conveyed to his audience Socrates' desire to control the story. So, yes, I understand and accept that one of the attractions of climatism is that it offers a simple, persuasive and directive story about climate change.

But my response to this point is twofold. First, stories – even appealing ones – can lead us in dangerous or unhealthy directions. Stories have power.[16] They can sometimes hinder people's behaviour, distort moral or cognitive reasoning, or else cultivate undesirable emotional states – as in the case of TikTok host Charles McBryde. At the very least, we need to reflect carefully on the stories we live by and on who is controlling them. (It will be the task of Chapter 5 to point out some of these dangers.) My second point is related. Maybe after such reflection we might recognize other stories that bring us closer to understanding the wicked nature of the problem of climate change, stories that might be more inclusive and open-ended. (I will seek to offer some pointers in Chapter 6.)

My argument, ultimately, is that there is *no* single story that does justice to the complexities, paradoxes and dilemmas exposed by the reality of a changing climate of largely human making. Climate change means different things to different people. For many people, those meanings change over their lifetime and through different life experiences. This leads some scholars to refer to the 'plasticity' of the idea of climate change, a plasticity which suggests something that is almost the opposite of a fixed ideology.

Summary

In its totalizing ambition, climatism has recognizable similarities to other appealing ideologies, such as Marxism, liberalism or capitalism. It seeks to reveal and explain the true state of things and to provide orientation for political action in the present and the future. In demands total allegiance from its followers. The features of attraction I have emphasized here can be found to greater or lesser degrees in other ideologies, but their combination in climatism helps explain this ideology's unique and wide appeal.

Climatism offers a master-narrative which brings clarity and simplicity to a complex world, and one that rests on 'special knowledge' drawn from the sciences and social sciences. It frequently adopts an apocalyptic pose through its use of visual and verbal rhetoric. By seeing the world through a Manichean lens it provides a framework for moral sorting which has intuitive appeal. Although painting the picture a little too starkly, the American campaigner Michael Shellenberger captures something of the allure of climatism which I have described in this chapter:

> [It] is powerful because it has emerged as the alternative religion for supposedly secular people, providing many of

> the same psychological benefits as traditional faith. It offers a purpose – to save the world from climate change – and a story that casts the alarmists as heroes. And it provides a way for them to find meaning in their lives – while retaining the illusion that they are people of science and reason, not superstition and fantasy.[17]

It is important to point out that the impetus behind climatism is quite different from that which motivated the rise of organized climate change denialism in the 1990s and 2000s. This latter was driven by the defensive strategy of powerful fossil-fuel actors, who sought to defend their interests by discrediting the science upon which concerns about climate change were seen to rest.[18] As this chapter has shown, the allure of climatism is different and more varied. Rather than being orchestrated by a powerful cabal defending entrenched political power, the politics of climatism are more diffuse. On the one hand there is climatism's psycho-cultural appeal for mass audiences – one might almost say its populist appeal, in particular its totalizing and simplifying master-narrative. On the other hand there are the specific beneficiaries of climatism. These include political authorities, from local through to national scales, who use climate change explanations for 'things going wrong' to mask their own deficiencies of negligence or bad management; civic actors who find that showing the climate change visiting card gains them access to specific financial and political resources for use in their cause; and some scientific experts who find that their public status and influence is sustained through the ideology of climatism, as explored in Chapter 3.

It is now time to examine more directly some of the dangers of the ideology of climatism. As I made clear earlier, ideologies are not innately wrong or necessarily pernicious. After all, they are essential for structuring our interpretation of the world and they can guide political action. If the ideology of climatism posed no dangers, then my argument thus far might be interesting, but not particularly important. But the next chapter seeks to persuade you that there are some real dangers with this way of thinking. Reducing the politics of the present and future to the single objective of stopping climate change might be a bad idea.

5

Why is Climatism Dangerous?

The Narrowing of Political Vision

The island of Sumatra in Indonesia has lost nearly 50 per cent of its tropical rainforest in the last thirty-five years. One of the recent drivers of this loss is the growth of the oil palm plantation. Journalist Tom Knudson tells of the effect of these plantations on the Indigenous Kubu forest dwellers, living in the swampy areas of southeastern Sumatra:

> Even if new planting were stopped tomorrow, it would be too late for the Kubu people I met, whose once-rich rain forest pantry has been stripped bare by an oil palm plantation. 'I have lost my garden', a Kubu woman named Anna told me. 'I cannot grow the rubber, bananas, chilies, and other things I need to feed my family.' Not long ago, Anna said, plantation workers even bulldozed Kubu homes to plant oil palm. 'We tried to stop them', she said. 'We started crying. But the man said, "Keep quiet or I'll take you to the police."'[1]

The problem for people such as the Kubu is that Indigenous land claims are not legally recognized. Plantations of oil palms squeeze out local populations who have no recourse to the courts. They lose their independent livelihoods and become 'slave workers' for the oil palm industry.

One of the chief reasons for the huge growth of oil palm plantations this century in places like Sumatra has been new climate policies introduced by western nations. Notable among these is the EU's Directive on 'the Promotion and use of Biofuels and other Renewable Fuels for Transport', first introduced in 2003. Although targets for biofuel use were initially voluntary and non-binding, pressure from European climate campaigners and NGOs led to them becoming mandatory in the subsequent Renewable Energy Directive (RED) passed in 2010. This legislation stipulated that, by the year 2020, 10 per cent of road fuel for each EU Member State should be derived from renewable sources including, but not limited to, biofuels.

This well-meaning, but narrowly focused, climate policy greatly drove up demand for cheap crop-based biodiesel, suitably provided by palm and soy oil, much of which was sourced from southeast Asia and South America. It is estimated that since 2010 this policy has wiped out forests the size of the Netherlands. People such as Anna have been dispossessed of their land and livelihood, and habitats of rare species have been destroyed. And even the carbon dioxide emissions avoided by displacing fossil fuel-derived diesel with biodiesel have more than likely been offset by the *increased* emissions resulting from the rainforest clearance that ensued. Laura Buffet, energy director at the pressure group Transport & Environment, explains:

> Ten years of this 'green' fuels law and what have we got to show for it? Rampant deforestation, habitats wiped out and worse emissions than if we had used polluting diesel instead. A policy that was supposed to save the planet is

> actually trashing it. We cannot afford another decade of this failed policy. We need to break the biofuels monopoly in renewable transport and put electricity at the centre of the RED instead.[2]

Although the EU *did* subsequently introduce a ban on the use of palm oil for biofuels by 2021, and agreed to end the use of crop-based biofuels by 2030, the damage had been done. The UN's special rapporteur on the right to food, Jean Ziegler, condemned the growing of biofuels as 'a crime against humanity' because they diverted arable land to the production of crops which were then burned for fuel instead of sold for food.[3]

The problem of biofuels is that when assessing their desirability, many other issues beyond the potential benefits for climate must be considered. These include: the cost of production and their competitiveness with fossil fuels; their impacts on food, biodiversity, energy and water security and human health; employment provision and security; the sustainability of rural communities; and so on. A recent comprehensive review of the sustainability of biofuels concluded that estimates of their environmental impacts vary widely between studies. The uncertainties are huge. 'Biofuels do not exist in isolation', these authors concluded, 'but are part of much wider systems, including energy, agriculture and forestry. Like other production systems with which they interact, biofuels impact on various ecosystem services, such as land, water and food.' When biofuels are evaluated solely in terms of their benefits for the climate, without wider consideration of their impacts on other facets of sustainability, policies mandating biofuels will lead to perverse and misleading outcomes.[4]

The dangers of climatism

I use this example of EU biofuels policy to illustrate one of the inherent dangers of climatism: its unyielding determination to pursue a single goal without considering – or without *carefully* considering – the broader context of why climate change is a matter of concern in the first place. This is what author James C. Scott refers to in his classic book *Seeing Like a State* as 'a narrowing of vision'.[5] A narrow, selective, one-eyed view of reality brings many advantages, not least making a complex and messy world appear legible and manageable. It is attractive to state authorities. Reductionism and quantification give the appearance of a world that can be manipulated and controlled through rationally designed policies.[6] As explored in Chapter 2, this is what GDP has done for economic policy, hospitalization rates have done for Covid-19 policy, and global temperature and Net-Zero have done for climate policy.

This is part of the allure of climatism explored in the previous chapter. If stopping climate change is the only thing that matters – or the thing that matters above everything else – then the EU biofuels directive may make sense. For policy experts located in centres of power in Brussels or Washington, and who subscribe or succumb to the ideology of climatism, mandating the expanded production of biofuels to displace fossil carbon from the liquid fuel mix may appear to be a necessary and viable 'solution' to climate change. But climate change *isn't* everything. The social consequences of the EU's biofuels policy, especially for some of the poorest communities around the world and their quest for food security and livelihood sustainability, have been devastating. A policy designed to reduce the impacts of climate change fifty years from now has undermined the livelihoods of people and the habitats of species living *today*. And even the long-term climate benefits of biofuels are not secure. Many studies have suggested that the net effect on greenhouse gas emissions of biofuels produced from rapeseed, corn, sugar cane or palm oil has been marginal at best or negative at worst.[7]

As we have seen, climatism puts 'stopping climate change' at the centre of its call to arms and therefore leads to policy actions which promote this single goal. This 'narrowness of vision', this one-eyed view of the world, can result in perverse outcomes such as those triggered by the EU Biofuels Directive and the resulting dash for sourcing biofuels whatever the cost. But triggering perverse outcomes is only one of the dangers of climatism to be alert to. This chapter points to four others: mono-causal explanations, cultivating the discourse of scarcity, issue depoliticization, and anti-democratic impulses.

Mono-causal explanations

The ideology of climatism is always in danger of veering towards environmental determinism and is also at risk of embracing the more recent phenomenon of climate reductionism. I identified this tendency as Move 5 in Chapter 2, where it was pointed out that the effects that climate – and *changes* in climate – have on societies and ecologies are not determined by climate alone. Nor in many cases are they even mainly determined by climate. Mono-causal climatic explanations of past or future events underweight the many other factors which condition climate's influence on societies. At the same time, they downgrade the human adaptive response to climatic events. We are not passive bystanders. The effects of a changing climate are always mediated through a complex matrix of biological, social, cultural, technological, political and historical factors. Hurricanes can be deadly or not, heatwaves of 35°C can kill or not, tidal surges are devastating or not, depending on these contextual factors. Yes, it is important and valuable to be able to predict the course of future climate. But it is just as important to understand the many changing interactions between climatic and non-climatic factors which condition exactly *how* and *why* climates impact societies the way they do, both now and into the future.

So, climatism might be misguided. It might clearly be *wrong* to offer mono-causal climatic explanations for many disastrous events. But why might such explanations also be *dangerous*?

Let's return to the example of the Syrian civil war with which I started this book. This civil war was much more rooted in Syria's long history of ethnic tensions and political grievances exacerbated by President Assad's economic and social policies than it was caused by the effects of a human-induced drought. As the political scientist Jan Selby has shown, the narrative of a climate change-induced war was seeded in part by the Syrian regime itself, seeking to deflect attention away from its responsibility for a political crisis of its own making. Selby shows how one set of interests – namely those of President Bashar al-Assad's regime – was pivotal in persuading many western commentators and policymakers of the climatic origins of the civil war.[8]

Although the Assad government initially refused to recognize the depth of the agrarian crisis in northeast Syria, it eventually realized that placing the blame for the crisis on a drought 'beyond our powers' would be politically astute. The drought was 'beyond our capacity as a country to deal with', claimed the Minister of Agriculture. 'Syria could have achieved [its] goals pertaining to unemployment, poverty and growth if it was not for the drought', declared Deputy Prime Minister Abdullah al-Dardari. Selby points out that the Belgium-based International Crisis Group reported that the Assad regime would regularly take diplomats to the northeast of the country and tell them, 'it all has to do with global warming'. What was essentially a state-induced socio-ecological crisis became blamed on climatic processes taking place far outside the jurisdiction of Assad's regime.

For those influenced by the climatist ideology, this argument fell on welcome ears. A chain of causation leading from global emissions of greenhouse gases and ending with the horrors of the Syrian civil war was exactly what some commentators with a climate-shaped field of view were expecting to find. Blaming the climate for all sorts of ills has long been an attractive political strategy, well used over the centuries by imperial powers, colonial authorities and authoritarian states. For those predisposed to hear them, plausible monocausal climatic explanations can be offered to account for any number of ecological disasters or political failures. These range from unsustainable agricultural expansion, to failed vanity mega-projects such as dam building or groundnut plantations, to arguments against the redistribution of scarce resources to oppressed and minority groups.[9] Such reductive explanations are dangerous. The effects of believing them can excuse violent, wilful or negligent political misconduct. They can legitimate so-labelled 'climate' policies which, if they fail to recognize the interplay between multiple social, economic and technological factors and the competing political interests at stake, are ineffective at best and regressive at worst. Climate change *isn't* everything. Believing otherwise is to misread the complex and contingent relationships between climates and the societies that live with them.

Discourse of scarcity

A second danger of climatism is that it lends weight to the politics of scarcity. Climatism too readily adopts – or at least too often fails to challenge – apparently naturalized deadlines by which certain enumerated and scientized climate targets 'must' be achieved. This is what I drew attention to as Move 10 in Chapter 2. Future climate is understood only in terms of a threshold, a 'point of no return', after which political action becomes 'too late'. Rather than a comet hurtling towards Earth to bring about its total demise – as in the film *Don't Look Up* – the scarcity discourse inverts the image. Earth is imagined to be hurtling towards a climatic cliff-edge after which it is game over.

The recurring trope of 'time is short' feeds this discourse of scarcity in the public politics of climate change. Headlines such as 'Climate scientists to world: We have only 20 years before there's no turning back' and 'Climate change: 12 years to save the planet? Make that 18 months' are commonplace. Even the more sober language emanating from social scientists reporting on the IPCC's 2022 Report adopts this tone of scarcity: 'The window of opportunity for climate adaptation action is closing fast because of warming and development trends' and 'The window of opportunity is closing, rapidly. Time is running short.'[10]

Framing climate change through this lens of time scarcity results in two understandable, but dangerous, outcomes: the creation of ticking climate clocks and declarations of climate emergency. Both are now ubiquitous. One of the more prominent climate clocks is organized by the American-based Action Network and is located in New York.[11] Nearly 25 metres wide and towering above New York's Union Square, the clock counts down 'the critical time window remaining for humanity to act to save itself and its only home from the ravages of climate chaos'. The giant clock seems to demand that 'one looks up'. As I write, it displays the time to 'climate chaos' to be 6 years, 345 days, 12 hours, 56 minutes and 10 seconds. 'There is still time to avert disaster' it tells me, but only if we take immediate action. If we fail, we pass 'a point of no return that science tells us is likely to make the worst climate impacts inevitable'.

If indeed the end is imminent, then it is not surprising that declarations of climate emergency follow. Since 2016 there has been a growing number of jurisdictions and institutions declaring a climate emergency – from municipal authorities and nation states to universities and churches. In 2020, the UN Secretary-General offered an invitation to the world's nations to declare their own state of emergency: 'Can anybody still deny that we are facing a dramatic emergency? That is why today, I call on all leaders worldwide to declare a State of Climate Emergency in their countries until carbon neutrality is reached.'[12] Given the prevailing energy infrastructure, and the high carbon dependency of most countries' economies, this probably means licensing presidents and parliaments, dictators and despots, to operate within a 'state of emergency' for at least the next three decades.

The discursive construction of time scarcity and the consequent emergency politics around climate change have significant, and worrying, *political* and *psychological* implications. If time is short, then any action will do, so long as it reduces emissions. Everything must be conducted 'in a hurry'. Such scarcity leads to narrow-thinking, depoliticized techno-managerialism and psychological anxiety. Not surprisingly, making time appear scarce leads to short-term thinking. But 'doing whatever it takes' without consideration of wider consequences has dangerous political ramifications, some of which we will examine in more detail in the next section.[13]

Here, I point to the psychological effects of time scarcity. The effect of climate deadlines is to inhibit the human cognitive capacity to imagine a future beyond the stated deadline. The future is closed down to a 'point of no return' that, if crossed, is declared to be the end. For the Union Square climate clock, this 'end' is 22 July 2029. The construction of such a precise and inviolable ending is intended by climatism to inspire urgency. Too often, it simply induces emotions of fear – fear and disengagement.[14] We saw in the previous chapter how TikTok host Charles McBryde admitted to feelings of anxiety and depression about climate change. He is just one of many young people who feel the same. A 2021 survey of 10,000 people aged between sixteen and twenty-five across ten countries revealed the depth of anxiety many young people are feeling. Nearly 60 per cent of respondents said they felt 'very worried' or 'extremely worried' about climate change, and three-quarters said they thought the future was frightening. Two-thirds reported feeling sad, afraid and anxious. Over half agreed with McBryde in thinking that humanity is doomed.[15]

In addition to cultivating anxiety and fear amongst young people, there is a very real danger that having too many 'last chances' to effectively tackle climate change stifles attempts to enact real change. A few examples suffice. In October 2006, Tony Blair, the then UK Prime Minister, claimed: 'We have a window of only 10–15 years to take the steps we need to avoid crossing catastrophic tipping points.' This window has now closed. A decade later, in 2017, climate scientist Hans Joachim Schellnhuber declared the decisive year to be 2020: 'The climate math is brutally clear: While the world can't be healed within the next few years, it may be fatally wounded by negligence [by] 2020.' This moment, too, has come and gone. And just two years later, in July 2019, the heir to the British throne and well-known conservationist, Prince Charles, reckoned the decisive window to be eighteen months: 'I am firmly of the view that the next 18 months will decide our ability to keep climate change to survivable levels and to restore nature to the equilibrium we need for our survival.' For Blair, Schellnhuber and Prince Charles alike we are now in 'dead time'. Our fate is sealed, it is too late.

The psychological effect of deadlines that are repeatedly missed, and hence either continually extended or else new ones created, is likely to be either cynicism, despair or apathy. There are only so many 'moments of decision' that are believable, after which they become unbelieved. They then lose whatever galvanizing force for political action they may have had. By elevating 'stopping climate change' to be the one issue by which success is measured, climatism feeds time scarcity and therefore fuels these unhelpful psychological outcomes. Its effects on politics and on democracy itself can also be chilling, as I explore next.

Depoliticizing the issue

Taking its cue from the condition of time scarcity, a third danger of climatism is that it strangles public politics. With time always short, action is always urgent; the time to act is always now. There is not enough time for societies to reflect, deliberate or experiment. This creates discursive conditions that some scholars describe as *depoliticalizing*. It may seem paradoxical that one of the dangers of climatism is that it *de*politicizes climate change. Surely, one might say, by drawing attention to the importance and urgency of climate change, climatism seeks to do just the opposite. It reinforces climate change *as* a political issue. But we need to understand what is meant by depoliticizing.

Depoliticizing of issues happens when public debates are closed down and when alternatives are suppressed. It is fuelled by claims such as 'there is no alternative' or 'this is an emergency'. When issues become depoliticized, states frequently mobilize scientists, economists, technocrats and government agencies – even the military – to design and deliver predetermined outcomes, foreclosing wider debates about ends and means. We certainly saw this happen with national responses to Covid-19 in 2020. Under such conditions, matters of public concern are deemed to have become 'post-political'; that is, they are beyond the reach of open political argument and contestation. An emasculated form of 'politics' ends up simply being an instrument to deliver the pre-defined goal. An issue can be said to have become depoliticized when, for example, a society or a government claims that 'the science leaves no alternatives' or to be 'following the science' or 'listening only to the scientists'.[16]

It is in this sense then that one of the dangers of climatism is that it depoliticizes climate change. Aided by the discourse of time scarcity, the 'politics' of climate change is reduced to simply delivering the technical and economic mechanics of securing Net-Zero emissions by a given date. The values, choices and trade-offs underlying these mechanics remain protected from public view and scrutiny. Claiming that 'stopping climate change' is the concern above all others, 'because the science tells us so', closes down the possibility of advocating, or even debating, other policy goals. 'The end justifies the means', the motto of all totalitarian projects. The danger then becomes that climatism squeezes out from public politics other traditional and important political values, such as liberty, equality, pluralism and self-determination.

Anti-democratic impulses

This leads to a related danger. It is not just that climatism depoliticizes the issue of climate change. It also endangers fundamental democratic values. It is only a short step from failing to recognize, and thus to debate, competing values and interests with regards to managing the risks of climate change, to actively suppressing or censoring any public voice that challenges the dominant position. The danger here is that dissent is closed down. Climatism is totalizing not just in the sense of seeking to envelop all matters of public concern within a single master-narrative. This is what we saw in Chapter 4. Climatism also, at its most extreme, seeks to police the boundaries of what can and cannot be said about climate change, and by whom.

The delegitimizing of publicly expressed views might start off with the intention to target disingenuous claims that seek to dismantle or discredit scientific evidence about the realities of climate change. There may be some merit in this.[17] But this impulse easily extends its reach to policing statements not about the science of climate change, but about alternative climate policies or about valid political trade-offs. 'Climate denialism' starts as a label to describe flagrant dismissals of well-established scientific knowledge. But it can become one that is used to denigrate those who question any element of climatism's master-narrative or to impugn the motives of those who question the efficacy of favoured climate policy proposals. By following this path, climatism is in danger of feeding illiberal and anti-democratic impulses, of censoring critical but legitimate and minority voices in a polity.

We saw one example of soft censorship in Chapter 3, with the BBC's Bitesize website. Another example would be the use of the term 'climate delayer' to delegitimate voices which raise valid questions about the wider social impacts of the pursuit of Net-Zero targets, or which question the ability of renewable energy technologies to displace all fossil fuels. One egregious case of this latter was in a commentary published in *The Guardian* newspaper, shortly after the 2015 Paris Agreement was signed. The commentary was titled, 'There is a new form of climate denialism to look out for – so don't celebrate yet'.[18] By using the 'denialism' label, the author – a prominent US historian of science – delegitimated the views of several highly respected scientists who were arguing for an expansion of nuclear power to accelerate the move away from fossil fuels.

Yet democracies require dissent if they are to remain democracies. For dissent to be legitimated – and for debate to be genuine – ideological and value pluralism has to be tolerated, and even encouraged. Michael Freeden's account of the role of ideology in a society again recognizes this. To shut down dissent against a dominant ideology is to elevate that ideology to hegemonic status, something inimical to democracies. It runs counter to the multiplicity of ideological standpoints desirable, and unavoidable, in a free society. 'For dissent to be legitimated', says Freeden, 'and in order for debate to be pluralist, reasonable ideological disagreement has to be accepted as normal and permissible by the public at large.'[19]

Two specific examples of this impulse to censor can illustrate the wider point. In 2019, a dozen Democratic senators in the United States put forward a bill that would have placed legal constraints on certain uses of federal research funds. The proposed bill sought to outlaw the allocation of public funds to activities that might 'challenge the scientific consensus on climate change'. The bill failed, but the intention to suppress criticism and challenge was clear. Also in the United States, recent Twitter guidelines prohibit ads that might seem to contradict the conclusions of the IPCC. Commenting on both these moves to silence climate dissent, Roger Pielke Jr points out that such regulatory efforts to 'defend' established science from its critics are wrongheaded.[20] They fail to understand the nature of scientific inquiry and, indeed, the nature of scientific consensus. Silencing dissenting voices is not the way to establish the epistemic credentials or the public acceptance of climate science. Neither scientific knowledge nor scientific consensus represents inviolable eternal truth. Knowledge and consensus are forged through challenge, critique, disagreement and compromise. As argued in Chapter 3, climatists should not be seeking to prohibit such practices. Rather just the opposite. Open challenges to science can at times reinforce the credibility and importance of consensus understandings; and in some cases they may lead to the evolution or refinement of knowledge as new research is produced.

In similar – if more extreme – vein was a proposal from Catriona McKinnon, a UK political scientist, that international criminal law should be expanded to include a new criminal offence she calls 'postericide'.[21] By this

McKinnon means any 'intentional or reckless conduct fit to bring about the extinction of humanity'. Writing in 2019, she says 'that we should look for postericidal conduct' in the domain of reckless public behaviour or statements that would make climate change worse. She imagines scenarios where such a case of 'postericide' might be brought to an international criminal tribunal. 'Climate crimes against posterity' might not result from the direct actions of any single individual. But McKinnon argues that individuals who have responsibility over large corporations, or even nation states, can commit such climate crimes. As examples of such criminal behaviour she offers CEOs withholding information about the progressive impacts of climate change because it threatens their corporation's bottom line. Or presidents withdrawing an entire state from a global agreement on climate mitigation.

I offer this latter example as an extreme case of where climatism might lead. Such international law would place climate denialism on a par with holocaust denial. A proposal such as this is very unlikely to gain legal traction, not least because a much tighter definition of 'climate denial' would be needed. Being charged with criminal behaviour simply because one advocates for nuclear power, as in the example alluded to above, would be laughed out of court.

Perverse outcomes

I conclude this chapter by returning to the concern I started with. If climate change is believed to be everything, then policies thus inspired can easily lead to perverse outcomes. Policies designed solely with the intention of reducing human forcing of the climate system – so-called mitigation policies – often create new social, ecological and political problems of their own. Similar perverse outcomes, called mal-adaptation, may arise from pursuing poorly thought-out climate adaptation policies. These characteristics are archetypical of wicked problems – problems where whatever you do in response will create more problems[22] – of which climate change is perhaps the most emblematic example. In addition to the case of biofuel mandates discussed earlier, I offer some further examples.

During the 2010s, Germany embarked on an ambitious and aggressive energy transition – its 'energiewende' – to accelerate its move away from fossil fuel-based technologies. The nation closed down coal and gas generating capacity, and for other, non-climatic, reasons began to rapidly dismantle its nuclear power industry. Its massive investment in renewables was, however, unable to deliver sufficient replacement reliable production capacity within the necessary time-frame. The consequence was that during this decade Germany's dependence on Russian oil and gas grew substantially. By 2021, 60 per cent of German gas imports came from Russia, roughly double the proportion of 2010. (To varying degrees, other EU nations also increased their dependency on Russian oil and gas during this period.)

One may debate the extent to which Germany's green energy policies empowered Putin to pursue his long-standing goal of invading Ukraine. But there is no doubt that following his invasion in February 2022, Europe has faced an unprecedented energy security crisis. By pursuing an aggressive, unbalanced and too rapid energy decarbonization to meet climate mitigation targets, Germany became hostage to Putin's goodwill to keep the gas flowing. The irony is that to defuse this medium-term crisis, Germany had to restart some of its mothballed coal-fired power plants – thereby again increasing greenhouse gas emissions – and also extend the life of its three remaining nuclear plants that were rashly slated to close in 2022.[23]

Pursuing the decarbonization of economies through a narrow climatist lens creates another potential danger. The rapid decommissioning of fossil-fuel assets – for example coal mines – can easily fail to recognize the severe human costs that result. These costs in many cases are borne by some of the more marginal and precarious communities – for example Indian, Polish or South African coal miners – and include rising local unemployment, regional economic decline and growing social inequality. Creating such 'left-behind' communities as a result of a rapid energy transition raises the question of social justice. Why should it be these workers who bear a disproportionate burden of the welfare costs of the energy transition? More widely, the consequences of climatism may also sow the seeds of future political disillusionment, polarization and unrest. It is too easy in the 'dash to decarbonize' to be blind to the unequal effects and burdens that result. Transitions to cleaner energies need to be 'just' as much as they need to be rapid.

A different example of how perverse outcomes may result from too close an adherence to the demands of climatism concerns the case of solar climate engineering. This was briefly mentioned in Chapter 1 in the context of the aggressive pursuit of techno-fixes. The idea of injecting particulate matter in the stratosphere, in order to create, in effect, an artificial thermostat for the planet, is driven by climatism's narrow focus on controlling global temperature as the pre-eminent sign of successful climate policy. This particular climate change 'solution' has gained growing numbers of advocates in recent years. Yet implementing this speculative technology carries significant risks, even were it to be successful in its sole aim of lowering global temperature. Reviewing the evidence for the possible unintended consequences of such a technology, social scientist Aarti Gupta concludes that, 'such technological adventurism will also likely exacerbate inequalities and worsen, rather than mitigate, the long-term consequences of the climate crisis'.[24]

Perverse outcomes can also result from policy actions which seek to enhance the adaptative capacity of systems or societies to changing climate risks. As with interventions designed to reduce the magnitude of climate change, if adaptation is pursued through the narrow field of view that climatism too frequently promotes, then more harm than good can result. Such unintended outcomes are often described as 'mal-adaptation', by which is meant 'action taken ostensibly to avoid or reduce vulnerability to climate change that impacts adversely on, or increases the vulnerability of other systems, sectors or social groups'. Mal-adaptation manifests in various ways. It might increase emissions of greenhouse gases thus contributing to the very problem it is trying to adapt to; it might disproportionately burden the most vulnerable; it might increase exposure to climatic hazards, the so-called moral hazard; or it might lock in development pathways that limit the choices available to future generations. For example, a new sea wall in Fiji was constructed to protect communities from rising tides and storm surges. The engineers, however, failed to allow for stormwater drainage on the landward side of the wall, causing settlements to flood. Increasing irrigation in dryland regions to compensate for reduced rainfall can lead to the salinization of groundwater or the degradation of wetlands. Resettlement of small island state inhabitants can *reduce* their resilience to climate change, due to consequential unemployment, homelessness, landlessness, food insecurity, social marginalization, reduced access to common-property resources, and increased morbidity.[25]

RETORT

Climate Change Still Matters

The argument developed in this chapter should not be interpreted as dismissing either the reality or the importance of climate change for human affairs. There are many ways in which it matters very greatly. Social life and ecological systems are profoundly interdependent with climatic conditions, and changes in these conditions inevitably present new challenges for societies and ecosystems. Limiting the rate and magnitude of climatic change is a desirable and necessary ambition. Neither am I saying here that every policy measure that has been implemented over the past quarter century in response to climate change is dangerous, misguided or has yielded perverse outcomes.

However, my complaint against climatism is twofold. First, I am drawing attention to the dangers associated with what James C. Scott calls a 'narrowing of vision', something that the ideology of climatism easily leads to. It 'flattens' the world and reduces understanding of the future to a single dimension: the fate of global climate and the achievement of Net-Zero. It offers a selective view of reality in which the broader context of human development, political freedom, technological innovation, human adaptation and ecological evolution is marginalized or lost sight of. Second, climatism is in danger of constraining the open-ended working of politics. It offers only scientized and reductionistic understandings of the relationship between climate, society and policy. This outlook fuels, and in turn is fuelled by, the conviction that scientific knowledge provides an adequate basis upon which schemes of social manipulation and systems of geophysical control can be designed and implemented. Yet schemes of such ambition are political through and through. The values underpinning them must be exposed and opened to wider public scrutiny and contestation. The rhetoric of time scarcity should not be used to short-circuit democratic processes.

Summary

In this chapter I have alerted readers to some of the (potential) dangers that emerge from too aggressive a hold on the public and political imagination of the ideology of climatism. These dangers are inter-related in complex ways. But all of them can be traced back to climatism's beliefs that stopping climate change is the political project that trumps all others – that the end justifies the means – and that this goal must be achieved within a rapidly diminishing time window. I have pointed out the dangers of promoting mono-causal climatic explanations for many of the world's ecological and political disasters. And I have highlighted how climatism's appropriation of time scarcity creates a set of inter-related psychological and political risks: widespread eco-anxiety, fear and disillusionment about the future, especially amongst teenagers and young adults; the suppression of open societal debate about competing values and goals, turning climate change into a post-political issue; and the illiberal and anti-democratic impulses that too easily follow from a zealous conviction that 'there is no alternative'. Finally, I have drawn attention to the many ways in which the single-minded pursuit of narrowly drawn climate-related policy objectives can lead to perverse outcomes.

So far I have called attention to and defined the emerging ideology of climatism and explained how it is underpinned by distinctive scientific and social scientific claims. I have cautioned against the sciences and social sciences slipping too uncritically into climatist thinking. I have pointed to the allure of climatism's master-narrative, with its totalizing, gnostic, apocalyptic and Manichean instincts. And I have identified several of the most pressing dangers of succumbing to this ideology, including its narrowing of political vision which, often, leads to perverse outcomes from well-intentioned policies.

But is there an alternative? Have we gone too far down the road of embedding climatist thinking in our public speech, politics and policies to find another path? If not, what other possibilities are there for taking seriously the realities and challenges of a changing climate, but without falling into the alluring trap of climatism? In the next chapter I offer a number of correctives to the ideology of climatism which might help avoid some of its most dangerous features.

6

If Not Climatism, Then What?

Wicked Problems Need Clumsy Solutions

Hundreds of millions of households in poorer nations currently rely for indoor cooking on either open wood fires or stoves fuelled by kerosene, coal, animal waste or other forms of biomass. These households represent about 2.5 billion people, about 30 per cent of the world's population. The World Health Organization estimates that about 3.8 million premature deaths occur annually linked to illnesses associated with the household air pollution such cooking technologies create. These deaths occur predominantly among women and children, girls especially, who tend to do most of the cooking and household chores. Furthermore, continued use of these fuel sources also impacts upon women's and girls' security, income and education. For example, gathering and collecting these fuels, whether wood or cow pats, takes time that girls could otherwise spend in school.

One of the most affordable, clean and scalable solutions for replacing these dangerous and burdensome traditional technologies is liquid petroleum gas (LPG) cooking stoves. As a fuel, LPG emits about 33 per cent less carbon dioxide than coal per joule of energy and about 12 per cent less than oil. Furthermore, LPG emits almost no black carbon, likely the second biggest contributor after carbon dioxide to global warming. It also substantially reduces the deadly indoor smoke which kills millions through pneumonia, lung cancer and chronic obstructive pulmonary disease.

One of the seventeen Sustainable Development Goals (SDGs) agreed by the United Nations in 2015 is SDG#7, which is to 'Ensure access to affordable, reliable, sustainable and modern energy for all'. About $4.5 billion of investment is needed annually to secure universal access to affordable, reliable, sustainable cooking fuels such as LPG, only about 3 per cent of which is currently offered by richer countries. And to make matters worse, several European nations such as Germany and Norway are now seeking to ban financial investments in fossil-fuel projects in low- and middle-income countries. Such a ban would include investments in technologies such as LPG cooking stoves.

By fixating on one policy goal – that of 'stopping climate change' – the ideology of climatism is losing sight of much wider and more diverse sets of welfare goals and ethical imperatives. In a remarkable editorial, written for the leading science journal *Nature*, the development economist and policy advocate Vijaya Ramachandran recently denounced the effects of this proposed ban in stark words:

> This puritanical, one-size fits-all approach is bad for the climate and overwhelmingly leaves women breathing in dangerous smoke from dirty cooking fuels . . . Policymakers from rich countries might say they support women's empowerment, but to me they seem more interested in simplistic climate mitigation – and coercing smaller nations to make cuts and compromises – than in improving the lives of poorer women. The irony is that clean cooking fuels

> [such as LPG] are much better for the environment than standard fuels . . . Pious, performative, broad-brush bans on fossil fuels help no one. A more intelligent, data-led approach is needed to better protect the climate alongside vulnerable people in developing nations.[1]

We can see how this narrowing of vision arose by tracking back through some of the international agreements agreed under the auspices of the United Nations. In 1992, the world's national governments drafted and signed the UN Framework Convention on Climate Change. Article 2 of this Convention set out its 'ultimate objective':

> stabilization of greenhouse gas concentrations in the atmosphere at a level that would prevent dangerous anthropogenic interference with the climate system. Such a level should be achieved within a time-frame sufficient to allow ecosystems to adapt naturally to climate change, to ensure that food production is not threatened, and to enable economic development to proceed in a sustainable manner.

Notice here the important qualifiers about multiple objectives *beyond* the stabilization of greenhouse gas concentrations. The broader dimensions of sustainability – ecological, social, economic – were recognized as constraints on a too aggressive or single-eyed focus on climatic goals. These broader goals were then enshrined initially in the Millennium Development Goals agreed in 2000 and, subsequently, in New York in September 2015, when all UN Member States adopted the 2030 Agenda for Sustainable Development. This Agenda

> provided a shared blueprint for peace and prosperity for people and the planet, now and into the future. At its heart are the 17 Sustainable Development Goals . . . which are an urgent call for action by all countries – developed and developing – in a global partnership. They recognize that

> ending poverty and other deprivations must go hand-in-hand with strategies that improve health and education, reduce inequality, and spur economic growth – all while tackling climate change and working to preserve our oceans and forests.

Just three months later, however, in December 2015, the Paris Agreement on Climate Change, under the negotiating apparatus of the UNFCCC, was adopted by these same nations. Again Article 2 – but now within the Paris Agreement – elaborated the ultimate objective:

> This Agreement, in enhancing the implementation of the Convention, including its objective, aims to strengthen the global response to the threat of climate change, in the context of sustainable development and efforts to eradicate poverty, including by holding the increase in the global average temperature to well below 2°C above pre-industrial levels and to pursue efforts to limit the temperature increase to 1.5°C above pre-industrial levels, recognizing that this would significantly reduce the risks and impacts of climate change.

Yes, the Paris Agreement *recognizes* 'the context of sustainable development' and 'efforts to eradicate poverty', but it effectively trumped the SDGs, signed just three months earlier. The goal of securing global temperature within a certain numerical range took precedence over a broader set of welfare ambitions, in part because of the success of climate scientists and government negotiators in characterizing the goal of climate policies in terms of a single, and seemingly simple, index. This is what I described as Move 3 in Chapter 2.

Antidotes against climatism

Climatism's narrowing of vision leads to the type of short-sighted and regressive policy that Ramachandran above is railing against – a ban on investments in fossil fuels in poorer countries. Although motivated by a virtuous goal – slowing the rate of global warming by reducing greenhouse gas emissions – the policy contributes to the continued discrimination against women and children in these countries.

How can this thinking be subverted? What is the alternative to climatism? It needs saying that ideologies rarely get overturned or dismantled in a hurry. But there are a number of arguments that can be made which might help loosen the grip that climatism has on much of today's political and moral imagination. In broad terms, these arguments follow the general principles of *political pragmatism*, which values pluralism over universalism, flexibility over rigidity, and practical results over utopian ideals.[2] When applied to climate change, such pragmatism calls for widening the field of view when thinking about and developing policy measures for enhancing human and ecological welfare. It recognizes that there are legitimate competing human values, interests and aspirations beyond delivering Net-Zero by a given date, and that policy trade-offs both within and between nations must be acknowledged, debated and navigated.

In this chapter I offer five antidotes – five ways of thinking about the world and its future in relation to a changing climate – that counteract the dangers of the narrow climatist thinking identified in the previous chapter. These correctives are to: foreground scientific uncertainty, defuse deadline-ism, promote 'technologies of humility', recognize the plurality of values, and pursue a plurality of goals. These correctives have some overlap with the antidotes to the emerging 'climate logic' identified by Aykut and Maertens and listed in Table 1 in Chapter 1: namely, plural ways of knowing, alternative globalities, participatory futuring, social transformations.

Foreground uncertainties about the climatic future

It is necessary to take more seriously the uncertainties embedded in all climate predictions and in all projected future impacts that are derived from them. These uncertainties are endemic because all models are fallible, as climate modellers well know.[3] But uncertainties also arise because all model predictions are conditional upon the pathways of future human development, which are inherently unpredictable. The projected impacts of these climate predictions need to make heroic assumptions about all the other dimensions of change that will condition how future climate risks are co-constructed alongside social changes. Or else they simply ignore these other changes that will occur in the human world, as in climate reductionism. Climate models are not truth machines and future scenarios are not crystal balls which allow one to see into the future. The IPCC does a good job of systematically assessing and summarizing the latest state of the climate sciences and social sciences with regards to these questions. But indeterminacy, uncertainty and ignorance about the future abound and persist.

Climatism rests on too confident a view of what can be known of the future. It is likely to err on the side of putting too much faith in scientific knowledge and the numbers it generates about the future. Climate scientists and modellers will continue to claim that ever more powerful models will be able to simulate with ever increasing accuracy and precision the future outcomes of complex interdependencies between physical, ecological, social and technological systems. This was Move 6 as described in Chapter 2. And it is exactly what the EU's Horizon Europe research programme promises to deliver by 2030: 'a "full" digital replica of Earth ... a highly accurate digital model of the Earth to monitor and predict the interaction between natural phenomena and human activities'.[4]

The first step in dismantling climatism is to treat such claims with great scepticism. The sciences and the social sciences are only able – and always will only be able – to see the future 'through a glass darkly'. Adaptation decisions are better made as hedges against a range of uncertain futures than as attempts to optimize based on uncertain predictions. Because of this lack of foreknowledge, policymakers need to know when to look beyond science and embrace other forms of analysis, reflection, wisdom and judgement. Framing, informing and guiding decisions about future policy requires much more than science. By itself, scientific knowledge offers no moral vision, no ethical stance and no political architecture for delivering the sort of worlds that people desire.

'Technologies of humility'

A second antidote to the dangers of climatism follows directly from this. It is to adopt what science studies scholar Sheila Jasanoff has called 'technologies of humility'.[5] By this she means 'disciplined methods to accommodate the partiality of scientific knowledge and to act under irredeemable uncertainty'. In other words, she urges that in the face of the unknown future humility should replace hubris. This is a broader argument than the one above about recognizing the specific limitations of climate prediction and the projection of climate impacts. It is rather about recognizing the *general* limits of human knowledge and foresight. It is about acknowledging the unforeseen contingencies of the future – whether pandemics, military or cyber wars between states, global economic recessions, failed states – and hence the inability of strategic planning to deliver precise policy targets ten or more years into the future. As one commentator put it: 'The present is a muddle. The future is an even bigger muddle whose basic coordinates we cannot predict, let alone control.'[6] This antidote is in effect about dismantling climatism's hubris.

What needs dismantling are the elements of the climatic logic referred to in Chapter 1, the logic of a globalized, panoptic, scientized control of the future. There is no omniscient mind – or mind collective – sitting in the cockpit of Earth directing matters towards a globally planned benign outcome, with success measured according to whether a single numerical target for the planet is secured. In contrast, the reality is that societies struggle along, reacting to problems as they emerge with incremental changes, making many missteps along the way. Ambitious visions to manage the Earth's physical and social complexity – whether using the promise of the EU's Destination Earth, machine learning or artificial intelligence – are chimeras. And a humbler disposition towards the future and towards systems of control would recognize that scientific knowledge is only one input into wise and just decision-making.

Take the example of social tipping points. Their existence is sometimes assumed presumptuously to suggest that policy levers can be identified and used to purposefully nudge entire societies into large-scale social transformation. Such faith in these levers rests on knowing enough about human behaviours to be able to control them, and on social and cultural worlds being predictable. Yet knowledge of social tipping points is rudimentary at best, even if it is possible at all. There is little evidence to suggest that we are even close to being able to advise policymakers how to 'tip' complex socio-technical systems into channels of behaviour that deliver predetermined outcomes. The inability of epidemiological models of the Covid-19 pandemic to capture the complex social and cultural dynamics of human behaviour – as opposed to the relatively predictable microbiology of the virus – bears witness to the need for such caution.[7]

Another example of the hubris of climatism which such humility undermines is the legal case brought against the UK government in 2022 by three non-profit organizations, ClientEarth, Friends of the Earth and the Good Law Project.[8] These plaintiffs argued in the High Court of Justice in London that the UK's strategy for reaching Net-Zero emissions by 2050 was in breach of the statutory 2008 Climate Change Act because it failed to set out by exactly how much individual policies will cut emissions. In response, government lawyers agreed that its strategy would only achieve 95 per cent of the legally mandated target by the interim date of 2035, the shortfall amounting to about 10 million tonnes of carbon dioxide. This legal wrangling over 10 million tonnes of carbon dioxide (the UK's current emissions are about 340 million tonnes) projected thirteen years into the future reminds one of arcane theological arguments from the early modern period about how many angels can dance on the head of a pin. It was a pedantic legal case driven by climatism's over-reliance on precise numerical projections and the illusion of control.

Such a court case may have symbolic political value, but it is of no practical value. Who could believe that *any* government's foresight of all relevant future contingencies – wars, pandemics, recessions, inflation, technological changes, shifts in cultural values – could possibly yield this degree of precise control over the future. For example, the UK's carbon emissions *increased* by 20 million tonnes between 2020 and 2021, for just these reasons. It is the direction of travel that matters far more than knowing or guaranteeing the precise destination. With regards to climate change, this destination is so far in the future, and is so characterized by deep uncertainty, that we cannot possibly imagine it. Even harder is to *control engineer* the path projected backwards from some desired state of the economy or planet in that unimaginable future. Under such conditions of deep uncertainty, delivering a national economy within a 5 per cent margin on such a long-term goal amidst all the muddle would in fact be a remarkable achievement.

There are no 'cliff-edges'

A third antidote is to recognize that it is unlikely that there are decisive cliff-edges in the climate system waiting for the human world to fall over. And even if there are, it is even less likely that we can find out pre-emptively *where* they are. 'Tipping points' are useful metaphors to aid our thinking about non-linear changes in physical systems. But they are metaphors, not to be taken literally.[9] Conversely, any deadlines that are set concerning dates by when certain things 'must' be achieved need to be recognized as *human creations*. They are not commands spoken to us from the external physical world. If climatism is to be dismantled, the tyranny of deadline-ism – and its inducements to emotions of failure, cynicism, apathy or fear – must be defused. Rather than imagining cliff-edges or crevasses, a better metaphor to think with is the picture of a gradual incline. Every 0.1°C of global warming that occurs increases some of the risks of climate change; with every 0.1°C of global warming avoided some of those risks are reduced. This antidote to climatism defuses the psychological fear and paralysis which distorts or inhibits creative human action in the world.

A good way to think about this is to adopt the framework proposed by journalist Max Roser: 'The world is awful. The world is much better. The world can be much better.'[10] These three statements do not contradict each other. As Roser explains, 'We need to see that they are all true to see that a better world is possible.' Roser applies his thinking to the case of child mortality. In today's world, 5.9 million children under the age of fifteen die every year; that is 4.3 per cent of all children. *The world is awful.* However, about 150 years ago, roughly 50 per cent of all children died 'prematurely'. *The world is much better.* And Roser uses the case of the European Union where today only about 0.45 per cent of all children die before age fifteen to argue that, if this is true in Europe, then in principle it is possible elsewhere also. *The world can be much better.*

Applying this heuristic to climate change moves us away from cliff-edges and deadlines. We live neither in the best of all possible climates, nor in the worst. In England and Wales, on average, around 800 excess deaths are caused each year due to heat, a number vastly outweighed by the approximately 60,000 excess deaths caused by cold.[11] Flooding in Pakistan in both 2010 and 2022 killed between 1,500 and 2,000 people, and the European heatwave of 2022 caused about 20,000 premature deaths. *Climate fatalities are large.* Droughts in

China and south Asia in the twentieth century killed millions. Within living human memory, tropical cyclones in the Bay of Bengal killed hundreds of thousands and flooding along southern North Sea coasts killed thousands. Fatalities from such climatic disasters have been reduced by at least an order of magnitude owing to improved forecasting and early warning, better infrastructure and more efficient management systems. As societies become more prosperous, they are better able to cope with weather extremes. *Climate fatalities are much reduced.* But there is more that can be done. Improved land-use planning and zoning, investments in more adaptative and smart infrastructure, and real-time public communication alerts to vulnerable communities using smartphone technologies can further reduce future fatalities from climatic hazards, even as they change in their intensity and geographical range. *Climate fatalities can be reduced further.*

This moves us beyond the mindset of time scarcity, endings and doomism, and instead adopts the language of possibility and emancipation. Above all, it undermines the message to teenagers and young adults that their generation is doomed and will have no future. Instead, it offers a message of hope, that their lives, taken in a collective aggregate, can be better than that of their parents and grandparents. Yes, these grandparents and parents – and we ourselves – have set in motion this ongoing change in the climate. But it is possible that human ingenuity and effort *can* limit the extent of future warming and *can* develop new technologies and strategies to adapt to the changes that result. Rather than repeating messages of failure and endings to teenagers and young adults, the alternative to climatism should motivate young people to contribute to a future world that 'can be much better'.

Plurality of values

This then leads to my fourth antidote to climatism, which is to give space to the diversity of political values and individual preferences within and between different polities. As soon as one recognizes that 'what is at stake with climate change is not human extinction, nor civilisational collapse, nor billions of unnecessary deaths'[12] – the comet is *not* approaching Earth – then one can see that there are legitimate and competing human values and political trade-offs that must be navigated when designing responses to climate change. Since the challenges raised by climate change are ultimately about values, it is essential to recognize value pluralism.

The idea of 'value pluralism' is a long-standing idea in moral and political philosophy, given great impetus by Isaiah Berlin's 1969 book *Four Essays on Liberty*. Berlin recognized that choosing between 'competing goods' was an inevitable feature of being human. The important point for my purpose is that the values that individuals or collectives hold with respect to what to do about climate change are neither singular nor infinite. They are plural. For example, how we balance the interests of future generations against those of today, whether we prefer centralized or decentralized forms of governance, or how we balance risk-taking versus precautionary styles of management . . . these are all value-laden judgements which vary widely. The different values that actors bring to these questions are rarely, if ever, in harmony with each other or even commensurable.[13]

The ideology of climatism too readily assumes that there *is* – or, more accurately, that there *should be* – a single and universal narrative which can bind together human knowledge, morality and strategy. Take these three frequently heard pronouncements: 'The science is clear: climate change is real', 'Climate change is the greatest challenge facing humanity' and 'Climate change is a moral issue'. The ideology of climatism desires to stitch together these three elemental claims into a single unifying master-narrative, as we saw in Chapter 4. A good example of such an aspiration comes from the collective thinking of three UK authors: a defence expert, an energy strategist and a systems analyst. These authors believe that tackling climate change requires the design of a single universal strategic narrative – 'a dynamic and persuasive system of stories, organically generated and encouraged between government, business and civil society' – that will unite and motivate all relevant audiences worldwide. Such a unifying narrative would bring together the diversity of literature and projects into a cohesive, coordinated and effective message. Furthermore, they believe that 'a government-led, iterative process of narrative forming should commence, incorporating as wide a range of stakeholders as possible. The outcome of this iterative process should be a short, digestible and persuasive narrative that is then naturally propagated by those stakeholders.'[14]

The possibility of such a vision is entirely at odds with value pluralism, let alone with the geopolitical realities of a fractured world.

The opposite of such climatist thinking is the idea of 'clumsy solutions' which emerged in the 1990s as the corollary of wicked problems. Wicked problems do not have a single strategic narrative nor do they lend themselves to solutions based on a set of converging values. They cannot be resolved by more empirical research or by an optimizing rationality. Even less so in the case of wicked problems such as climate change that transcend national jurisdictions and that mobilize multiple conflicting political interests. Clumsiness – and its close associate pragmatism – recognizes the multiple subjectivities involved in identifying and framing wicked problems like climate change and the inevitability of the plural values that political subjects and institutions will bring to the issue.[15]

The consequence of value pluralism is that climate change becomes just one problem among many. Different collective interests driven by diverse values will see different priorities: child health, poverty eradication, ending hunger, gender equality, economic stability, livelihood autonomy, conserving marine life, and so on, as captured by the SDGs. The danger with climatism is that it runs roughshod over the different values that lie behind these preferences. Climatism pronounces that 'stopping climate change' takes precedence over all other issues; the antidote is to recognize this diversity of legitimate concerns, value traditions and political priorities. It is to explicitly facilitate such arguments globally and without necessarily seeking consensus solutions. More important than working towards a 'single strategic climate narrative' is to find ways of working with disagreement, seeking out shared procedures for how to deal with value pluralism.

Plurality of goals[16]

Recognizing and respecting value pluralism leads to a fifth antidote to climatism: diversifying policy goals away from securing abstract metrics and scientized global proxies such as global temperature or Net-Zero emissions. Value pluralism demands goal pluralism. The antidote to climatism is to design, promote and mobilize around diverse policy goals that have a direct bearing on locally contextualized and specific social-ecological welfare outcomes.

Take the example of forests. In 2021, at COP26 in Glasgow, driven by the rush to demonstrate commitment to Net-Zero ambitions, a wide coalition of nations pledged $12 billion over the next four years to halt deforestation, with a focus on the tropics and sub-tropics. But forests are not only sinks for carbon dioxide. And 'halting deforestation' not only protects a carbon sink. It has other effects as well. For many communities in the Global South, the harvesting of wood from forests provides their income and their energy. There are many different types of low-intensity wood harvesting practices that are essential for the world's poor, and yet which maintain a functioning forest ecosystem. Poorly designed, institutionally ineffective and scientifically uninformed bans on such harvesting practices simply drive them 'underground'. The livelihoods of many vulnerable people are made illegal overnight. For the sake of the world's poor and their forests, 'climate science and policy must not forget these people in the race to reduce emissions from deforestation'.[17]

The climate pragmatism I am advocating takes shape under the umbrella of the Sustainable Development Goals. In terms of international agreements and political legitimation, the SDGs carry the same force and urgency as the Paris Agreement on Climate Change. A total of 193 nations have adopted Agenda 2030, which commits them to delivering the SDGs by 2030. Foregrounding the multiple welfare and ecological goals represented by the SDGs in any response to climate change more easily allows both policy synergies and political trade-offs to be explored than does a single focus on delivering Net-Zero. It facilitates the expression and negotiation of different political priorities and cultural values, and it enables political processes to work creatively towards a wider range of multiscale, partial and pragmatic policies.

Working with value pluralism and with the diversity of legitimate political concerns represented by the SDGs, climate pragmatism offers a different vision of the challenges facing the world because of human-caused climate change. It offers a more robust strategy than the ideology of climatism because it allows and finds ways of aligning different and otherwise mal-aligned political interests. Climate pragmatism is also more robust because it is supported by multiple justifications for action. It does not rest on scientific predictions of the climate future, or on apocalyptic rhetoric about world endings, or on a Manichean worldview of two warring tribes.

Take the example of decarbonizing energy systems. Climate pragmatism turns it from a single-focus issue – 'stopping climate change' – into a many-focus issue.[18] Decarbonization is about much more than simply reducing carbon dioxide emissions. It is about improving health outcomes for many vulnerable groups, enhancing energy security – both national and local, and widening the range of ownership arrangements of energy services. This perspective has been well articulated by Yemi Osinbajo, the Vice-President of Nigeria, a country of 220 million people, projected to reach 400 million by the year 2050. Osinbajo argues that any global transition away from carbon-based fuels must account for the economic differences between countries:

> For countries such as my own, Nigeria, which is rich in natural resources but still energy poor, the transition must not come at the expense of affordable and reliable energy for people, cities, and industry. To the contrary, it must be inclusive, equitable, and just – which means preserving the right to sustainable development and poverty eradication, as enshrined in global treaties such as the 2015 Paris

> climate accord. Nigeria and other African countries are committed to a net-zero future . . . But our commitment to climate action cannot be separated from our energy needs. A just global energy transition must include Africa, and it cannot deny our people their right to a more prosperous future.[19]

Osinbajo goes on to challenge a narrow climatist view of decarbonization. Instead of hampering Africa's economic development, the rich world should help the continent's energy producers 'secure financing for vital natural gas projects that can serve as a bridge to net-zero and for renewable projects and the modern grids required to handle them'.

Australian cultural geographer Carol Farbotko offers another example. Fetishizing the reduction of carbon emissions without a sufficiently broad concern for social justice issues – such as exemplified in SDG#1 (ending poverty), SDG#6 (clean water and sanitation), SDG#8 (securing decent work) or SDG#10 (reducing inequality) – risks further endangering those who are economically poor or socially marginalized. Such people, Farbotko writes, are 'already facing danger from both financial and climate risks. Their newest risk is the risk of being excluded from, or only superficially included in, the emerging risk calculus of the climate-finance meta-system.'[20]

RETORT

Climate Change isn't Everything

One obvious criticism of the argument put forward in this chapter is that what I call 'climate pragmatism' – for example, prioritizing the SDGs over Net-Zero – won't necessarily deliver the objective of the Paris Agreement: stabilizing global temperature at between 1.5° and 2°C. This may be the case. But this gets to the heart of my complaint against climatism. Climate change isn't everything. It is quite easy to imagine future worlds in which global temperature exceeds 2°C warming which are 'better' for human well-being, political stability and ecological integrity, for example, than other worlds in which – by all means and at all costs – global temperature was stabilized at 1.5°C.

It is inevitable that there will be trade-offs between different SDGs and between the SDGs and stabilizing global temperature. Not everything can be secured and certainly not everything can be secured simultaneously. Win-win solutions can only go so far. Several of the SDGs – for example eradicating poverty (SDG#1), securing quality education (SDG#4), ensuring decent work and economic growth (SDG#8) – will require the expansion of affordable and reliable energy services for billions of people, not least in Africa and south Asia. Only some of these services can be delivered by zero-carbon energy, as Yemi Osinbajo observes.

The problem with climatism is that exploring these trade-offs is short-circuited by placing global temperature above all other goals. Even debating such trade-offs is seen by some as defeatist. 'Stopping climate change must be the project we pursue above all others.' The Paris Agreement merely 'recognizes' the context of sustainable development and efforts to eradicate poverty, but is very clear about what constitutes success: stabilizing global temperature increase at between 1.5° and 2°C. I am arguing that we need a broader measure of success than that offered by climatism. Elevating the SDGs over and above the Paris climate goal is necessary. The SDGs decentre the science-based narratives of global temperature, carbon budgets and Net-Zero which often stand in as proxies for hidden or unspoken values. The thrust of climatism needs inverting: 'We should deliver the SDGs whilst recognizing the context of global warming.'

Summary

This chapter has outlined a different way of thinking about the significance of climate change for human affairs than is offered by the ideology of climatism. The five antidotes, or correctives, to climatism recognize the limits of Earth System prediction and of human knowledge about the future more generally. They challenge the illusion of systemic control of the Earth's, and humanity's, future. They defuse the paralysing imaginary of a looming cliff-edge of endings and its associated rhetoric of 'we only have so many years'. They supplant it with possibilities for more pragmatic and plural – and more focused – interventions in the world. They explicitly position the challenges of a changing climate, real as they are, and the important desire to rein in the scale of future changes in climate, as subservient to the need for sustainable human and ecological welfare as encapsulated by the Sustainable Development Goals. Taken together, these antidotes also better recognize the political and geopolitical realities of international negotiations and national decision-making than does the ideology of climatism.

To quote Otto von Bismarck: 'Politics is the art of the possible, the attainable – the art of the next best.' I believe climate pragmatism, not climatism, offers the best way for cultivating such an art.

7
Some Objections
'You Sound Just Like . . .'

The argument I have advanced to show that climate change isn't everything is likely to generate some strong reactions. Some people will use this book, and my challenge to the ideology of climatism, to try to discredit climate science as a whole. Others will find justification in what I have written for marginalizing or even eradicating concerns about climate change from political debates and from policy development. I thoroughly disagree with both these interpretations.

On the other hand, there will be those who see in what I have written traces of what some authors, led by ecological economist William Lamb, have recently called 'discourses of delay'.[1] Climate delay discourses, they suggest, have the intention to discourage 'climate action', where climate action appears to mean policies or campaigns for societal transformations that seek to mitigate climate change. The proponents of this idea group discourses of delay into one of four categories, those that either (i) redirect responsibility to others; (ii) push for non-transformative solutions; (iii) emphasize the downsides of climate policies; or else (iv) surrender to the inevitability of climate change. The suggestion of these authors is that people who fall into one or more of these positions should be labelled 'climate delayers' – a term maybe not as pejorative as 'climate deniers', but pejorative nonetheless. Climate delayers, they say, are people who 'argue for minimal [climate] action' and who draw attention to 'the negative social effects of climate policies and raise doubt that [climate] mitigation is possible'. They promote illegitimate political positions, ones that the public needs to be 'warned against' and which need to be 'overcome' using 'inoculation strategies that protect the public from their intended effects'.

With this in mind, it is likely that there will be some readers who place my foregoing argument in the 'climate delayer' category. They would accuse me of 'eroding public and political support for climate policies' and of disorienting and discouraging 'ambitious climate action'. They would perhaps see this book as 'misrepresenting the climate crisis' and of failing to communicate 'the dramatic pace of global warming [and] the gravity of its impacts'. Some might even go so far as to accuse me of the (speculative) international crime of 'postericide' (see Chapter 5).

By seeking to discredit in this way contributors to debates about how best to respond to climate change, Lamb and his co-authors reveal one of the alluring features of climatism that I identified in Chapter 4, namely its adoption of a Manichean worldview. 'You're either with us or you're a delayer.' Rather than recognizing that questions about responsibility, feasibility, efficacy, justice, radicalism vs incrementalism, coercion, 'free-riders', and so on, are important and necessary ones to raise in public debates about climate policies, they would seek to dismiss such concerns by blacklisting those who voice them. These authors therefore also reveal the depoliticizing and anti-democratic tendencies in climatism, to which I pointed in Chapter 5.

Through a series of short retorts, earlier chapters responded to some of the criticisms that might be directed at my argument against climatism. Now, in this final chapter, I respond more directly to some other anticipated criticisms of my position.

1. 'Climate science is not alarmist'

Complaint: 'You seem to suggest that the climate sciences and social sciences – perhaps even the IPCC as well – are alarmist, the implication being that predictions of climate change and projected climate impacts are exaggerated.'

No, I am not making this explicit claim. But I *am* urging some caution when interpreting scientific knowledge claims. In this respect, my argument against climatism is twofold. First, I am pointing to the inherent and inescapable uncertainties in all predictions about future climate and its possible impacts. This is well recognized by the IPCC which, over the years, has developed a carefully calibrated language to capture some of these uncertainties. For example, the most recent IPCC Report from 2021 on the science of climate change stated that continued Amazon deforestation increases the possibility that during the twenty-first century this ecosystem will cross a tipping point into a dry state. But this statement was deemed by the IPCC to have '*low confidence*'. Similarly, the IPCC stated that a collapse in the Gulf Stream – technically called the Atlantic Meridional Overturning Circulation – is deemed '*very unlikely*'. On the other hand, it is '*virtually certain*' that further decreases in Northern Hemisphere snow cover extent will occur as the climate warms, and it is known with '*high confidence*' that, for most regions, heatwaves will increase in frequency and intensity.

It is therefore essential to pay careful attention to the varying levels of scientific confidence about different facets of the changing climate system. Some changes are 'virtually certain', some are judged to have 'low confidence' and some are just 'not known'. Scientific inquiry is always 'feeling its way' towards a greater understanding of the physical world, but not in a linear path towards ever greater clarity and certainty. There are often zig-zags or U-turns where things are provisionally discovered, then re-analysed and subsequently modified, until something else is discovered which complicates the story or requires different questions to be answered. As scientific inquiry proceeds, rather than the future becoming ever clearer and more predictable, it may in fact appear to become more uncertain. As we know more, the plot thickens, so to speak. The case of the sources of methane illustrates this nicely. After levelling off in the 2000s, the concentration of methane in the atmosphere has accelerated in the last few years, but the likely and known sources of methane emissions can't account for this. Scientists have therefore sought to understand what might be behind the recent acceleration. This has led to different competing ideas – tropical wetlands, whether driven by higher rainfall or maybe by warmer temperatures, or reduced rates of chemical decomposition of methane in the air – but the issue as yet remains unresolved.

The second thing I am doing is drawing attention to something that is well established from sociological studies of all the sciences. Namely, that the questions scientists ask and the discoveries they make – and therefore the predictions they offer – are always conditioned by the political and cultural contexts in which they work. We remain ignorant of some things about the physical world because of the questions we *don't* ask. Scientific practice and scientific knowledge are thus exposed to a range of internal and external pressures to pursue personal or collective agendas. In historian of science Stephen Shapin's memorable phrase and eponymous 2010 book, science therefore is 'Never Pure'.[2] This is not to discredit science as a powerful method for understanding how the physical world works, but it warns that science, as a collective social enterprise, needs always to be alert to unduly distorting influences. This appreciation cautions against interpreting all claims emanating from science as literal truth: they require careful reading and critical inspection, and should only ever secure our provisional assent. Some have previously argued that these pressures are more likely to lead to 'climate science erring on the side of least drama' because of its innate conservatism.[3] I would argue that the opposite danger is at least just as likely, as evidenced in some of the examples offered in Chapter 3.

2. 'Climate change is an existential risk'

Complaint: 'If climate is a "condition of life", as you suggest, then surely we *should* put the goal of stopping the climate from changing above everything else. It *is* an existential danger to life.'

I recognize some of the voices who would regard climate change this way. For example, in 2018 the United Nations Secretary-General, António Guterres, made the bald claim that 'we face a direct existential threat' from climate change, while others such as Jem Bendell at the University of Cumbria in the UK point to a growing community of people who have concluded that 'we face inevitable near-term societal collapse', a community to which Bendell himself seems to belong. And then there is the American columnist Tom Engelhardt, who has placed humanity on suicide watch for itself. 'Even for an old man like me', he says, 'it's a terrifying thing to watch humanity make a decision, however inchoate, to essentially commit suicide.' Bill McGuire, in *Hothouse Earth: An Inhabitant's Guide*, seems to view climate breakdown as both inevitable and terrifying; those who would deny this he labels 'climate appeasers'. McGuire's is one of the most extreme expressions of climate doomism around – he *wants* his readers to be frightened – and it is hard to find much hope in the view of the future he offers.[4]

Yet I believe these prognostications of 'extinctionism' and societal collapse which pervade the discourse around climate change, and which feed climatism, are wrong. They are also counter-productive, as I have shown in Chapter 5. And they do a disservice to the development, justice, peace-making and humanitarian projects being undertaken around the world today. As I argued in Chapter 4, ongoing climate change is not like a comet approaching Earth. There is no good scientific or historical evidence that climate change will lead to human extinction or the collapse of human civilization. Climate is a changing condition to which all sentient life is continually adapting – yes, imperfectly and unevenly, and at different speeds, and with danger and death always lurking. But adaptation is the natural response of ecological systems to change, as much as it is of social systems. Corals, for example, adapt speedily to changing environmental conditions; genes that help a coral organism survive physically in a particular environment will likely be passed on to their offspring. Although the speed of ocean warming and acidification might challenge these natural adaptive processes, new genetic techniques make it possible to speed up naturally occurring evolutionary processes.[5] Social systems too are continually adapting to changing environmental conditions, for example through improved early warning systems, through innovative smart materials that can help keep buildings cool, or through new land-use zoning.

Again, don't misread me. Climate kills and climate change is real. The risks induced by a changing climate are serious. Efforts to mitigate these risks and to adapt to them are important. But climate change will not wipe out human life, let alone all life on the planet. And it is questionable whether annual deaths from climate change will ever exceed those resulting from non-communicable diseases such as heart or lung failure, dementia or stroke. Climate change is a significant risk with uneven effects, but it is not a collective existential one. It is a risk that needs to be attended to, but this must be done in the context of other present or emerging risks, such as nuclear war, pandemics, preventable childhood mortality, failed states, air pollution and so on. Climatism becomes a dangerous ideology when it feeds the belief that the human species faces extinction because of climate change and when it pronounces that there are only so many years left to stave off collapse – thereby claiming warrant for declarations of a perpetual climate emergency. This is dangerous talk that opens the door to misguided one-eyed techno-solutions, as we saw in Chapter 5.

3. 'Justice is much more central to climate change advocacy than you imply'

Complaint: 'Your criticism of so-called "climatism" seems to imply that climate campaigners are so obsessed with securing global temperature and delivering Net-Zero that they are completely blind to any broader questions of justice, equity, reparations and so on. It's *not* all about ends as you claim; it *is* about means also.'

It is true that many people campaigning for the need to stop climate change point to its complex political, social and historical causes. They frame climate change as a question of justice. For example, in her recent book *What Climate Justice Means and Why We Should Care*, political theorist Elizabeth Cripps[6] says that since climate change results from 'colonialism, slavery, oppression and systematic disregard for basic human rights', those responsible for such things have the greatest duty to act. But then many of these same voices gather behind the climatist announcement that stopping climate change is the most important challenge facing humanity. They take their cue from the climatist claim that the priority above all others is to deliver Net-Zero by a certain date in order to secure global temperature at a fixed level. Emphasizing this 'end' as the goal of climate action *does* marginalize scrutiny of the 'means' of getting there.

This tension was nicely articulated in a Reuters news report on the day after the IPCC published the Working Group 2 Report of their Sixth Assessment in April 2022.[7] Reuters drew attention to the requirement that efforts to reduce greenhouse gas emissions 'must be fair' and must account for 'countries' other key priorities – such as development in poorer nations'. If emissions are cut swiftly and deeply across economies 'at the expense of justice, of poverty eradication and the inclusion of people, then you're back at the starting block', said Fatima Denton, one of the IPCC Report's 278 authors.

This is exactly the point of my argument in Chapter 6, in which I seek to offer antidotes to the narrowing vision of climatism. The point is to recognize that, for many, 'stopping climate change' *isn't* the number one priority. The SDGs articulate many of these 'countries' other key priorities', as Denton describes them, such as poverty eradication, ending hunger, quality education, affordable and clean energy, and decent work and economic growth. As the Reuters report makes clear, the rush to put climate change ahead of all other issues inevitably downgrades the importance of those issues.

Climate justice campaigners should put aside the rhetorical claims of climatism, such as 'follow the science' or 'we must achieve Net-Zero by this date'. Instead, they should argue more directly for their goals of distributional justice, equitable access to resources, or securing well-being for all. If they do not, climatism will sweep away these concerns in the drive to 'hit the targets', whether by techno-fixes such as solar climate engineering or carbon capture and storage, or through regressive policies such as denying rural women in India access to clean fuels, or hampering economic development in Africa. As we saw in Chapter 5, climate justice will be brushed aside in the name of a 'higher good': namely, stopping climate change by whatever means. But climate justice is *precisely* about means; it is only incidentally about stabilizing weather systems. The climatist vision of 'stopping climate change' at all costs might turn out to be a dangerous ally in the campaign for social justice. It might turn out to have little interest in securing such objectives and might even get in the way of achieving them.[8]

4. 'If capitalist consumerism is to be challenged then maybe a counter-ideology such as climatism is needed'

Complaint: 'If climatism is indeed an ideology, then maybe such an ideology is exactly what is needed to mobilize people on the scale necessary to challenge environmentally destructive ideologies such as capitalism, nationalism or state socialism.'

This is very much the argument developed in Naomi Klein's book *This Changes Everything: Climate vs Capitalism*. Klein's argument – and that of others of like mind – is that the real significance of climate change is its signalling that capitalism is reaching its end. Or, at least, that it is necessary to ride the wave of climatism to ensure that through aggressive campaigning and discrediting of capitalism's social licence to operate it *does* reach its end. But it is a dangerous move to position the ideology of climatism as a direct opponent to the ideology of capitalism. I pointed to some of the reasons for this in Chapter 5. Any challenge to capitalism – or indeed to state socialism or any other environmentally destructive ideology – needs defending and advocating on explicitly political, moral and normative grounds. It should not rest its authority, credibility or legitimacy on the sciences.

Using climate science as a trump card to shut down ideological opponents by saying 'do what the science says' or 'follow the scientists' does two related and unwelcome things. It uses the claims of climate science as a substitute for the hard argumentative work of persuading one's opponents of the moral, economic or political superiority of one's ideology (see Chapter 3). And, consequentially, it thereby risks turning science into nothing more than another ideological weapon to be deployed in the political arena. This is not good for science and it is not good for politics. This was pointed out by several science-analysts and science communicators shortly after the April 2017 'March for Science' in Washington DC to protest against President Trump's ignoring of science. This politically charged event carried the danger that scientists would be perceived by the American public as just another interest group, driven more by ideology than by evidence. Science communicator Matthew Nisbet was particularly critical, stating that the march would only deepen 'partisan differences, while jeopardizing trust in the impartiality and credibility of scientists'.[9]

5. 'You sound just like a climate denier, a climate delayer or a Pollyanna, an excessively or blindly optimistic person'

The final complaint is self-evident. My response is to say, 'Don't be concerned with what or who I *sound like*, but engage directly with the concerns and arguments I have advanced.' In other words, don't try to force my position into some pre-established category, whether that be denier, delayer, contrarian, lukewarmist, or whatever else. I have made clear that my use of the term 'climatism' is radically different from Steve Goreham's in his 2010 book *Climatism!* (see Chapter 1). The argument made here is *my* argument, no one else's. The general point is that the positions on climate change being advanced by different authors should be evaluated and judged on their respective merits. One should not accept or reject a position 'just because' its author is labelled a 'climate alarmist' or a 'climate contrarian' or anything else.

Climate change is not a nail awaiting a hammer

Just over sixty years ago, in February 1962, the American philosopher Abraham Kaplan was invited to give the after-dinner speech at the end of a three-day conference of the American Educational Research Association. This was held in his home city of Los Angeles and, as reported later, 'the highlight of the 3-day meeting ... was Kaplan's comment on the choice of methods for research'.[10] Kaplan was concerned that scientists should be careful when selecting their methods for investigating research problems. The fact that certain methods happen to be at hand, or that a scientist happens to be conversant with a particular method, is no assurance that the chosen method will be appropriate for the presenting problem. He later formulated his thinking as Kaplan's Law of the Instrument: 'Give a small boy a hammer, and he will find that everything he encounters needs pounding.' In a later elaboration of this Law, Kaplan commented: 'We tend to formulate our problems in such a way as to make it seem that the solutions to those problems demand precisely what we already happen to have at hand.'

The idea Kaplan articulated in 1962 captured the imagination of many in his audience. Others picked up on his idea and began to generalize it. One wrote about 'the tendency of jobs to be adapted to tools, rather than adapting tools to jobs', while the psychologist Abraham Maslow condensed Kaplan's warning thus: 'I suppose it is tempting, if the only tool you have is a hammer, to treat everything as if it were a nail.'[11] The aphorism is now commonplace, almost banal. Even voicing it in its most reductive form – 'Think hammer and nails' – will often get the point across.

I think that climatism has some similarities with the behaviour to which Kaplan's Law of the Instrument points. Climatism approaches the world in such a way that every problem seems to be a result of the ongoing changes in climate being caused by human actions. If that is so, then the solution to these problems would appear to be arresting climate change. This effectively inverts Kaplan's Law. 'If everything looks like a nail then what we need is a hammer.' If everything looks to be a result of climate change, then clearly what we need to do is to stop climate change. The 'climate hammer' then gets reduced to three inter-related ambitions: to predict future climate with ever greater precision; to eliminate fossil fuels; to deliver a Net-Zero emissions society.

But these are very limited tools for meeting the world's ongoing social, development and environmental challenges. Take the case of the subsiding Mekong Delta in southeast Asia. The Delta is home to 17 million people and produces nearly 10 per cent of the world's rice. But it lies on average just one metre above sea level and is slowly subsiding. A climatist would approach this challenge through the lens of the ongoing changes in climate and sea level facing the region. For sure, there are changes afoot in the region's climate. And the sea level is rising, with projections of future increases this century of between 30 and 70 cm. For the climatist, the imperative in securing the sustainability of the Delta would be to arrest the rate of global climate change by doubling-down on global carbon dioxide and other emissions and to enhance adaptation to the predicted changes in climate.

But this is wrong. The problems facing the Mekong Delta are not a nail awaiting a hammer. The problems are not first and foremost caused by climate change. Dam-building starves the Delta of sediment. In-channel mining removes a further fifty-four megatons of sand from the Delta each year. And agricultural intensification and flood regulation have replaced natural waterways and mangroves with dikes and aquaculture. These problems are compounded by groundwater pumping for urban and agricultural use and by the spiralling lock-in of flooding leading to more dike construction which, by making river channels more rigid, leads to more flooding.[12]

The solutions to the Delta problem are therefore many-headed and need multiple tools. Yes, the sea level is rising and the climate is changing, but few of the needed solutions derive from the hammer of stopping climate change. The Mekong Delta is an intricate regional socio-ecological system needing a variety of carefully designed interventions to guarantee its sustainability. For example, further high-impact dam construction should be stopped; sediments should be directed through or around existing dams; riverbed sand-mining should be phased out; agriculture in the Delta needs transforming; the connectivity of its floodplains needs to be maintained; and nature-based coastal protection, such as through mangrove enhancement, needs incentivizing. The tools in the toolbox need to be diverse and utilized locally and in context.

Moving to a very different part of the world, the Arctic, we find the same danger of seeing a nail because we have a hammer. The challenges of life today in the 'cold north' are many. The issues that dominate discussions among Arctic residents themselves are these: the lived experiences of health and securing health care; high rates of suicide and substance abuse; low educational attainment; economic viability; sustaining culture and language; food security; the effects of extractive industries; and the day-to-day demands of existence in remote areas with extreme climates. As one group of Inuit residents and scholars explains, fixating on the changing climate of the Arctic – which is undoubtedly occurring – misses much that is pressing and worrisome:

> An outsized focus on climate alone generates certain types of mitigation agendas, namely reducing greenhouse gasses by nations and corporations in the Global North, rather than addressing multiple and intersecting issues related to health, poverty, education, economic viability, cultural vitality, and justice.[13]

For Arctic dwellers, as for those living in the Mekong Delta, a changing climate is but one challenge among the many they are facing. Using the hammer of climatism to meet these challenges can be seen as a latter-day form of colonialism: 'Focusing on research framed primarily around sea ice and climate change can actually obscure . . . community impacts, responses, meanings, politics and even hydrological processes.'[14] A more sensitive framing of many of the local and regional problems facing communities around the world would in fact *de-centre* climate change from the story, instead placing a changing climate within deeper historical and political contexts and alongside wider developmental needs. Rather than raising the battle-cry of climatism, 'We must stop climate change before everything else', this framing would allow for greater self-determination about what in fact needs to be done, when, how and by whom.

For the people of the Mekong Delta, and for those dwelling in the high Arctic, climate change matters, but it isn't everything. And this, I believe, is true for everyone. The challenges of a changing climate are real. But they can *only* be understood – indeed, they *must* be so understood if they are to be responded to appropriately – in the context of other presenting problems and in view of the many other prospective changes in the world that are underway.

The present isn't all about climate change, and the future mustn't be reduced to climate. Stopping climate change isn't the only thing that matters. Climate change *isn't* everything.

Further Reading

Beyond the citations provided to specific sources and examples used directly in the text, this final section points readers to a wider collection of books upon which I have drawn, to greater or lesser degrees, in the development of my argument. Some of these books I first encountered several years ago, others more recently. In their various ways they each explore one or more of the dimensions of climate change with which I grapple. To trace my own thinking on the questions addressed here, I draw attention to an earlier trilogy of books I have written on the science, politics and cultures of climate change, namely *Why We Disagree about Climate Change: Understanding Controversy, Inaction and Opportunity* (Cambridge University Press, 2009), *Weathered: Cultures of Climate* (SAGE, 2016) and *Climate Change: Key Ideas in Geography* (Routledge, 2021).

There are any number of books which explain or summarize the science behind climate change. Since I am not trying to repeat or condense this huge literature in the present book, I mention just three books about (climate) science which have particularly informed my case. The first is Tim Lewens' *The Meaning of Science* (Penguin,

2015), in which he explains in simple terms the nature of scientific knowledge and how it is made. Lewens shows persuasively why important questions about the purpose and meaning of science matter, explaining them in personal, practical and political ways. Second is a short book titled *The Rightful Place of Science: Politics* (Arizona State University, 2013), edited by Pascal Zachary. Zachary gathers together a collection of essays by American scholars who insist that scientific research and technological change cannot be understood separately from their socio-political contexts. Coming to terms with this contextual reality allows for a much richer understanding of the nature of science and scientific knowledge. On climate science more specifically is *The Climate Demon: Past, Present, and Future of Climate Prediction* (Cambridge University Press, 2021) by R. Saravanan. Saravanan, a modeller at Texas A&M University, provides an accessible overview of the history, strengths and limitations of climate modelling and prediction. He explores and illustrates the difficult challenges climate scientists face when communicating science's inherent uncertainties about its predictions of the future.

Central to my argument is a particular understanding of the relationship between expertise, politics and democracy. Four books that explore this in different ways have been very important for me. Robert Boyers' *The Tyranny of Virtue: Identity, the Academy, and the Hunt for Political Heresies* (Simon & Schuster, 2019) is a powerful defence of the liberal values of tolerance, dissent and argument against the tyranny of political conformity and the simplicities of moral sorting. Taylor Dotson's *The Divide: How Fanatical Certitude is Destroying Democracy* (MIT Press, 2021) offers a provocative defence of democratic pluralism, challenging the idea that some undeniable truth – whether involving an appeal to scientific facts, common sense, nature or the market – will resolve political disputes. Dotson draws in part upon the turbulent history of climate change in American political life. In a similar vein, Amanda Machin's *Negotiating Climate Change: Radical Democracy and the Illusion of Consensus* (Zed Books, 2013) shows why 'listening to the scientists' is so wrong in the case of climate change. What democracies need, she argues, are vigorous, honest and open debates not just about goals, but about the means to achieve those goals. Also recommended is Gil Eyal's *The Crisis of Expertise* (Polity Press, 2019). Eyal is professor of sociology at Columbia University and in this book he tackles the troubling rise in the 'mis-trust of experts'. He skilfully dissects some of the reasons for this 'mis-trust', not least being the growing tendency to try to settle political differences by using the authority of science coercively. This is exactly one of my complaints against climatism.

Climate change means different things to different people. Three books which explore the different social meanings of climate change in an accessible way are: Philip Smith and Nicholas Howe's *Climate Change as Social Drama: Global Warming in the Public Sphere* (Cambridge University Press, 2015), Candis Callison's *How Climate Change Comes to Matter: The Communal Life of Facts* (Duke University Press, 2014) and Michael Brüggemann and Simone Rödder's *Global Warming in Local Discourses: How Communities around the World Make Sense of Climate Change* (OpenBook Publishers, 2020). All three books give the lie to the belief that untying the Gordian knot of climate change rests with science, more science, better science, or more consensual science. What climate change means to different political actors and civic communities is rooted in divergent human beliefs, values and life experiences, which the ideology of climatism too easily ignores or passes over, or views as too difficult to deal with. Relatedly, there is the collection of case studies edited by Evan Berry, *Climate Politics and the Power of Religion* (Indiana University Press, 2022). Through a series of nine studies from around the world, Berry's contributors show that the lens of religion is far more important for shaping how communities respond to climate change than is adopting the ideology of climatism, which rests on a scientized and globalized view of the world.

The following three books are valuable for understanding the power of narratives and stories around climate change. In *Storylistening: Narrative Evidence and Public Reasoning* (Routledge, 2021), Sarah Dillon and Claire Craig radically expand the notion of what constitutes the 'evidence' that should inform policymaking: not just scientific facts and statistical models, but stories too. They make a powerful case for why narrative literacy is as important for wise and effective public decision-making as is scientific literacy or numeracy. More specifically on climate change there is Raul Lejano and Shardul Nero's *The Power of Narrative: Climate Skepticism and the Deconstruction of Science* (Oxford University Press, 2021). They provide important insights into the different rhetorical and semantic strategies that different sides in the climate change debate adopt in pursuing their political goals. The co-written volume *Climate Change Scepticism: A Transnational Ecocritical Analysis* (Bloomsbury Academic, 2019), by Greg Garrard, Axel Goodbody, George Handley and Stephanie Posthumus, investigates the cultures and rhetoric of climate sceptical discourses in the UK, Germany, the United States and France by analysing literary texts. This analysis reveals that climate scepticism is a many-headed phenomenon and points towards ways of overcoming partisan political paralysis.

The following two books explore in greater depth the problems of climate reductionism, and what I call mono-causal climatic explanations of disasters and other world events. Jan Selby, Gabrielle Daoust and Clemens Hoffmann's *Divided Environments: A Political Ecology of Climate Change, Water and Security* (Cambridge University Press, 2022) focuses on the links between climate change, water and security, particularly in the Middle East and North Africa. The authors show how the security implications of climate change are very different from how they are often imagined. They decisively expose the naivety and dangers of climate reductionism, so evident in popular accounts of climate conflicts and water wars. Roger Pielke Jr has been writing about disasters and climate for many years, and *The Rightful Place of Science: Disasters and Climate Change* (University of Arizona, 2018, 2nd edition) is an excellent short account of his well-documented and perceptive insights. Disasters are a serious problem, as are human-caused changes to our climate. Taking both disasters and climate change seriously, and addressing them effectively, requires the recognition that they are not serious for the same reasons. The pathways for addressing them are different and must respond to different information, arguments, motives and policies.

To help understand the process of 'climatization' there is *Globalising the Climate: COP21 and the Climatisation of Global Debates* (Routledge, 2017) by the political scientists Stefan Aykut, Jean Foyer and Edouard Morena. They show how several general global issues – such as development, energy, international security and migration – have become climatized in recent years, particularly during the COP21 negotiating conference held in Paris in 2015. Aykut and his co-authors identify some of the trends that I draw attention to in my book, but they stop short of calling out climatism as an ideology or explaining its full effects – and dangers – in today's world. For something a little closer to my own position on climate change there is Jon Symons' *Ecomodernism: Technology, Politics and the Climate Crisis* (Polity Press, 2019). Symons makes a reasoned case for ecomodernism, the belief that technological innovation and universal human development hold the keys to a just and environmentally sustainable future. Ecomodernism, says Symons, is not a reactionary position, but promotes a third way between laissez-faire economics and anti-capitalism. For understanding the basics of ideology, I have leant heavily on Michael Freeden's *Ideology: A Very Short Introduction* (Oxford University Press, 2013).

Finally, let me mention Lucian Boia's *The Weather in the Imagination* (Reaktion Books, 2005). I first came across this little book more than fifteen years ago and have re-read it a couple of times since. Although somewhat idiosyncratic, it offers a fascinating short history of the idea of climate change in human culture. Boia points to the long-standing cultural anxiety, traced back through human history, that climate is fragile and vulnerable and that, once destabilized, it will no longer offer amenable conditions for human life and flourishing. This anxiety remains with us today, but we should recognize it for what it is and not exaggerate it, let it paralyse us or allow it to narrow our vision of the future.

Notes

Introduction

1 This account is based on Marwa Daoudy's *The Origins of the Syrian Conflict: Climate Change and Human Security* (Cambridge: Cambridge University Press, 2020).

2 For a summary of these claims and their respective sources see: J. Selby, O.S. Dahi, C. Fröhlich and M. Hulme, Climate change and the Syrian civil war revisited: a rejoinder. *Political Geography* 60, 2017: 253–5.

3 I. Johnston, Climate change helped caused Brexit, says Al Gore. *The Independent*, 23 March 2017.

4 A. Stechemesser, L. Wenz et al., Strong increase of racist tweets outside of climate comfort zone in Europe. *Environmental Research Letters* 16, 2021: 114001.

5 Europe floods: Merkel shocked by 'surreal' devastation. *BBC News*, 18 July 2021, https://www.bbc.com/news/world-europe-57880729

6 M. Hulme, Reducing the future to climate: a story of climate determinism and reductionism. *Osiris* 26(1), 2011: 245–66.

7 T. Mitchell, *Prisoners of Geography: Ten Maps That*

Tell You Everything You Need to Know about Global Politics (London: Elliott & Thompson, 2015).

1 From Climate to Climatism

1 R. Fox, How climate change helped the Taliban win. *Reaction News*, 16 September 2021, https://reaction.life/how-climate-change-helped-the-taliban-win

2 One of the more interesting and readable accounts of how the science of global climate change developed during the past two centuries is Sarah Dry's *Waters of the World: The Story of the Scientists Who Unravelled the Mysteries of Our Seas, Glaciers and Atmosphere – and Made the Planet Whole* (Chicago: University of Chicago Press, 2019).

3 M. Hulme, *Why We Disagree about Climate Change: Understanding Controversy, Inaction and Opportunity* (Cambridge: Cambridge University Press, 2009); M. Hulme, *Weathered: Cultures of Climate* (London: SAGE, 2016); M. Hulme, *Climate Change: Key Ideas in Geography* (Abingdon: Routledge, 2021). Of this trilogy, *Weathered* is perhaps best read first since it takes the broadest historical perspective; *Climate Change* is in many senses an update of the thinking I first put forward in *Why We Disagree. . . .*

4 For high energy physics see: E. Gibney, What's the carbon footprint of a Higgs boson? It varies – a lot. *Nature* 611, 2022: 209, and for whales see: R. Chami et al., A strategy to protect whales can limit greenhouse gases and global warming. *International Monetary Fund*, December 2019, https://www.imf.org/Publications/fandd/issues/2019/12/natures-solution-to-climate-change-chami#author

5 For these latter points, see: M.A. Rajaeifar, Decarbonize the military – mandate emissions reporting. *Nature* 611, 2022: 29–32.

6 D. Jayaram, 'Climatizing' military strategy? A case study of the Indian armed forces. *International Politics* 58(4), 2021: 619–39.
7 L. Maertens, Climatizing the UN Security Council. *International Politics* 58, 2021: 640–60.
8 S. Grant, C.C. Tamason, P. Kjær and M. Jensen, Climatization: a critical perspective of framing disasters as climate change events. *Climate Risk Management* 10, 2015: 27–34.
9 S.J. Pyne, *The Pyrocene: How We Created an Age of Fire, and What Happens Next* (Oakland: University of California Press, 2022).
10 P. Jenkins, *Climate, Catastrophe and Faith: How Changes in Climate Drive Religious Upheaval* (Oxford: Oxford University Press, 2021).
11 Grant et al., Climatization. For *The Guardian* story see: J. Vidal, 'We have seen the enemy': Bangladesh's war against climate change, *Guardian*, 9 May 2012, https://www.theguardian.com/environment/2012/may/09/bangladesh-war-against-climate-change
12 This is a feature of climate change framing and discourse I discuss in: M. Hulme, Problems with making and governing global kinds of knowledge. *Global Environmental Change* 20(4), 2010: 558–64.
13 S.C. Aykut and L. Maertens, The climatization of global politics: introduction to the special issue. *International Politics* 58(4), 2021: 501–18.
14 S. Goreham, *Climatism! Science, Common Sense and the 21st Century's Hottest Topic* (New Lenox, IL: New Lenox Books, 2010), p. 161.
15 M. Freeden, *Ideology: A Very Short Introduction* (Oxford: Oxford University Press, 2003), p. 2.
16 Freeden, *Ideology*, p. 32.
17 There is also a similarity between ideologies and myths, if we think of the latter in the anthropological

and non-pejorative sense of the word; i.e., myths as 'stories that embody fundamental truths underlying our assumptions about everyday or scientific reality'.

18 G. Monbiot, *How to Stop the Planet Burning* (London: Allen Lane, 2006), p. 15.

19 For this quote and Kirk's resignation, see: HSBC banker quits over climate change furore. *Financial Times*, 7 July 2022, https://www.ft.com/content/5ff24114-5777-4d00-a014-ad36ce948d64 Stuart Kirk's original fifteen-minute talk is available on YouTube at https://www.youtube.com/watch?v=bfNamRmje-s

20 For criticism of Kirk's speech, see: G. Wagner, Climate risk is financial risk. *Science*, 10 June 2022, 1139. For examples of legitimate questions being raised by Kirk, see: J. Zammit-Lucia, Stuart Kirk, HSBC, and the politics of climate. *CEOWorld Magazine*, 10 June 2022, https://ceoworld.biz/2022/06/10/stuart-kirk-hsbc-and-the-politics-of-climate Also R. Pielke Jr, Why investors need not worry about climate risk? *The Honest Broker*, 23 May 2022, https://rogerpielkejr.substack.com/p/why-investors-need-not-worry-about

21 D. Mustafa, Pakistan must get rid of colonial mindset on water. *The Third Pole*, 9 September 2022, https://www.thethirdpole.net/en/livelihoods/opinion-pakistan-must-get-rid-of-colonial-mindset-on-water/?amp

22 M. Davis, *Late Victorian Holocausts: El Niño Famines and the Making of the Third World* (London: Verso, 2000).

23 A. Wijkman and L. Timberlake, *Natural Disasters: Acts of God or Acts of Man?* (London: Earthscan, 1984).

24 The 'lukewarmer' accepts that climate change is real and mostly man-made, but does not consider it to be a planetary emergency. For example, see: M. Ridley,

The Climate Wars and the Damage to Science (London: Global Warming Policy Institute, 2015). The case of the outright climate change denier would be represented by Figure 2, since they would not recognize that climate is in any way influenced by human activities.

2 How Did Climatism Arise?

1 S. Kuznets, *Response to Senate Resolution No. 220 (72nd Congress). A Report on National Income, 1929–1932* (Washington DC: Government Printing Office, 1934).
2 W.D. Nordhaus, *Can We Control Carbon Dioxide?* IIASA Working Paper, 1975. IIASA, Laxenburg, Austria: WP75063.
3 When Nordhaus was working on this problem in the mid-1970s, there was no generally accepted method for estimating global temperature and efforts at reconstructing the history of global temperature were still rudimentary, and somewhat contradictory. It was not at all self-evident that this was the 'correct' control variable to select.
4 D. Philipsen, *The Little Big Number: How GDP Came to Rule the World and What to Do about It* (Princeton: Princeton University Press, 2015).
5 Hulme, *Weathered: Cultures of Climate.*
6 W.B. Meyer, The perfectionists and the weather: the Oneida Community's quest for meteorological utopia 1848–1879. *Environmental History* 7(4), 2002: 589–610.
7 These ideas have been well developed in the work and writings of historical geographer David Livingstone. See for example: D.N. Livingstone, The climate of war: violence, warfare and climatic reductionism. *WIREs: Climate Change* 6(5), 2015: 437–44;

and D.N. Livingstone, Climate and civilization. In M. Boyden (ed.), *Climate and American Literature* (Cambridge: Cambridge University Press, 2021), pp. 58–74.

8 See the following, sometimes also referred to as 'The Bretherton Report', after the lead author, Francis Bretherton: National Research Council, *Earth System Science. Overview: A Program for Global Change* (Washington, DC: National Academies Press, 1986).

9 See for example: European Commission, *Destination Earth*, https://digital-strategy.ec.europa.eu/en/library/destination-earth The publicity brochure states: 'Destination Earth (DestinE) is a major initiative of the European Commission. It aims to develop a very high precision digital model of the Earth (a "digital twin") to monitor and predict environmental change and human impact to support sustainable development.'

10 See for example: S. Randalls, History of the 2°C climate target. *WIREs Climate Change* 1(4), 2010: 598–605.

11 See for example: R. Peet, The social origins of environmental determinism. *Annals of the Association of American Geographers* 75, 1985: 309–33.

12 A.C. Hill, COVID's lesson for climate research: go local. *Nature* 595, 2001: 9.

13 For the example of the Norwegian 'snowmen' see: J. Solli and M. Ryghaug, Assembling climate knowledge: the role of local expertise. *Nordic Journal of Science and Technology Studies* 2(2), 2014: 18–28. For a discussion of what good climate adaptation needs see: S. Dessai and M. Hulme, Does climate adaptation policy need probabilities? *Climate Policy* 4, 2004: 107–28.

14 J. Slingo et al., Ambitious partnership needed for reliable climate prediction. *Nature Climate Change* 12(6), 2002: 499–503.
15 For a good discussion of the history of the idea of the 'allowable' carbon budget see: B. Lahn, A history of the global carbon budget. *WIREs Climate Change* 11(3), 2020: e636.
16 An excellent account of the rapid institutionalization of Net-Zero as a policy goal is found in: H. Van Coppenolle, M. Blondeel and T. Van de Graff, Reframing the climate debate: the origins and diffusion of net zero pledges. *Global Policy*, 21 November 2022, https://onlinelibrary.wiley.com/doi/full/10.1111/1758-5899.13161?campaign=wolearlyview
17 See the argument put forward by M. Hulme, S.J. O'Neill and S. Dessai, Is weather event attribution necessary for adaptation funding? *Science* 334, 2011: 764–5.
18 See: S. Li and F. Otto, The role of human-induced climate change in heavy rainfall events such as the one associated with Typhoon Hagibis. *Climatic Change* 172, 2022: 7.
19 An excellent analysis of this problem is offered in Jesse Ribot, Violent silence: framing out social causes of climate-related crises. *The Journal of Peasant Studies* 49(4), 2022: 683–712.
20 S. Asayama, Threshold, budget and deadline: beyond the discourse of climate scarcity and control. *Climatic Change* 167(3), 2021: 1–16.
21 The most comprehensive description of the IPCC, how it functions and what influence it has is provided in: K. De Pryck and M. Hulme (eds.), *A Critical Assessment of the Intergovernmental Panel on Climate Change* (Cambridge: Cambridge University Press, 2022).

3 Are the Sciences Climatist?

1 Pielke, Why investors need not worry about climate risk?

2 The numbers in these scenario labels refer to the strength of the physical warming effect of the respective greenhouse gas concentration; RCP8.5 is therefore about 3.3 times the strength of RCP2.6. The precise explanation of these numbers need not concern us.

3 R. Pielke Jr and J. Ritchie, Distorting the view of our climate future: the misuse and abuse of climate pathways and scenarios. *Energy Research & Social Science* 72, 2021: 101890.

4 Z. Hausfather and G.P. Peters, Emissions – the 'business as usual' story is misleading. *Nature* 577, 2020: 618–20.

5 'We are concerned that . . .' is from Z. Hausfather et al., Climate simulations: recognise the 'hot model' problem. *Nature* 605, 2022: 26–9. The Voosen quotation is from P. Voosen, 'Hot' climate models exaggerate Earth impacts. *Science* 376, 2022: 685.

6 Hausfather et al. Climate simulations.

7 K. Brysse et al., Climate change prediction: erring on the side of least drama? *Global Environmental Change* 23(1), 2013: 327–37. Ultimately, the risks associated with climate change can only be expressed as subjective and conditional probabilities, and how people interpret and act on those risks is a matter of psychology, not of science. For example, see the discussion of 'fat-tailed distributions' in G. Wagner and M.L. Weitzman, *Climate Shock: The Economic Consequences of a Hotter Planet* (Princeton: Princeton University Press, 2015).

8 J. Loconte, One hundred years ago, 'following the science' meant supporting eugenics. *The Institute for Faith and Freedom*, 19 July 2022, https://www.faith

andfreedom.com/one-hundred-years-ago-following-the-science-meant-supporting-eugenics

9 For the points made in this paragraph, see: D. Roberts, *Fatal Invention: How Science, Politics, and Big Business Re-create Race in the Twenty-first Century* (New York: The New Press, 2012); A. Saini, *Superior: The Return of Race Science* (London: The Fourth Estate, 2019); J.L. Graves and A.H. Goodman, *Racism, Not Race: Answers to Frequently Asked Questions* (New York: Columbia University Press, 2021); J. Marks, *Is Science Racist? Debating Race* (Cambridge: Polity, 2017); and C.F. Lewis et al., Getting genetic ancestry right for science and society. *Science* 376, 2022: 250–2.

10 WHO, Climate change and health, 30 October 2021, https://www.who.int/news-room/fact-sheets/detail/climate-change-and-health

11 The report in full is: PBL, *Assessing an IPCC Assessment: An Analysis of Statements on Projected Regional Impacts in the 2007 Report*. The Hague/Bilthoven, Netherlands, 2010.

12 H. Horton, BBC removes Bitesize page on climate change 'benefits' after backlash. *BBC online*, 2 July 2021; https://www.theguardian.com/media/2021/jul/02/bbc-removes-bitesize-page-climate-change-benefits-backlash See also: https://www.bbc.co.uk/bitesize/guides/zcn6k7h/revision/5

13 S.H. Schneider, The 'double ethical-bind' pitfall. *Mediarology*, https://stephenschneider.stanford.edu/Mediarology/mediarology.html

14 See P. Kitcher, *Science in a Democratic Society* (New York: Prometheus Books, 2011), p. 184. I thank Mark Brown for this paraphrase of Kitcher's scenario.

15 For a discussion of this controversy and its significance see: M. Mahony, The predictive state: science, territory

and the future of the Indian climate. *Social Studies of Science* 44(1), 2014: 109–33.

16 For a good exposé of the dangers of statistical cherry-picking, see E.J. Wagenmakers, A. Sarafoglou and B. Aczel, One statistical analysis must not rule them all. *Nature* 605, 2022: 423–5.

17 An excellent account of this controversy is given in: F. Pearce, *The Climate Files: The Battle for the Truth about Global Warming* (London: Guardian Books, 2010). An analysis of the controversy which reveals the motivated reasoning of the scientists involved is found in: R. Grundmann, 'Climategate' and the scientific ethos. *Science, Technology & Human Values* 38(1), 2013: 67–93.

18 J. Watts, We have 12 years to limit climate change catastrophe, warns UN. *Guardian*, 8 October 2018, https://www.theguardian.com/environment/2018/oct/08/global-warming-must-not-exceed-15c-warns-landmark-un-report

19 See: S. Asayama, R. Bellamy, O. Geden, W. Pearce and M. Hulme, Why setting a climate deadline is dangerous. *Nature Climate Change* 9(8), 2019: 570–2.

20 B. Salter and C. Salter, Controlling new knowledge: genomic science, governance and the politics of bioinformatics. *Social Studies of Science* 47(2), 2017: 263–87 (p. 281).

21 G. Schmidt, Mmm-k scale climate models. *RealClimate*, 25 June 2022, https://www.realclimate.org/index.php/archives/2022/06/mmm-k-scale-climate-models/#ITEM-24379-0

22 J. Bohr, The 'climatism' cartel: why climate change deniers oppose market-based mitigation policy. *Environmental Politics* 25(5), 2016: 812–30.

23 For a good introduction to these questions, see: S. Sismondo, *An Introduction to Science and*

Technology Studies, 2nd edition (Hoboken, NJ: Wiley, 2009), and T. Lewens, *The Meaning of Science* (London: Penguin, 2015). Bruno Latour offers a more sophisticated analysis of the social nature of science in *Science in Action: How to Follow Scientists and Engineers through Society* (Cambridge, MA: Harvard University Press, 1988). For a robust defence of science and of how and why it remains trustworthy, see: N. Oreskes (ed.), *Why Trust Science?* (Princeton: Princeton University Press, 2019).

4 Why is Climatism So Alluring?

1 A. Trembath, '*Don't Look Up*' peddles climate catastrophism as a morality tale. *Foreign Policy Insider*, 18 December 2021, https://foreignpolicy.com/2021/12/18/dont-look-up-review-mckay-comet-climate-change
2 J. Stephens and R. McCallum, *Retelling Stories, Framing Culture: Traditional Story and Metanarratives in Children's Literature* (New York: Routledge, 1998), p. 6.
3 Potsdam Summer School, Towards a sustainable transformation – climate, energy and nature in a changing world (2022), https://potsdam-summer-school.org
4 N. Klein, *This Changes Everything: Capitalism vs the Climate* (New York: Simon & Schuster, 2014), pp. 7–8.
5 For Stenmark, the 'myth of the Absolute' is the belief that there is an Absolute outside of human history and independent of human limitations, one upon which we can base our judgement and actions and which makes it possible to avoid errors and mistakes. Myths of the Absolute are rooted in the experience of human finitude and fallibility and the realization (or fear) that nothing is certain. They are also connected to an experience of something beyond our limited

existence, something greater than ourselves. See: L.L. Stenmark, Storytelling and wicked problems: myths of the absolute and climate change. *Zygon* 50(4), 2015: 922–36.

6 For a good account of this see: N. Oreskes and E. Conway, *Merchants of Doubt: How a Handful of Scientists Obscured the Truth on Issues from Tobacco Smoke to Global Warming* (London: Bloomsbury, 2010).

7 For Thunberg on BBC radio, 23 April 2019, see: https://www.bbc.co.uk/news/av/uk-48018034 For testimony to the US House see: *NBC News*, 18 September 2019, https://www.nbcnews.com/science/environment/climate-activist-greta-thunberg-tells-congress-unite-behind-science-n1055851

8 I use the term 'apocalypse' here in the popular sense of cataclysmic destruction or endings, rather than its original religious meaning 'to disclose' or 'unveil'. For the quotation see: L. Buell, *The Environmental Imagination: Thoreau, Nature Writing, and the Formation of American Culture* (Cambridge, MA: Harvard University Press, 1995), p. 285.

9 G. Ereaut and N. Segnit, *Warm Words: How Are We Telling the Climate Story and Can We Tell It Better?* (London: Institute of Public Policy Research, 2006).

10 P. Barkham, 'We're doomed': Mayer Hillman on the climate reality no one else will dare mention. *Guardian*, 26 April 2018, https://www.theguardian.com/environment/2018/apr/26/were-doomed-mayer-hillman-on-the-climate-reality-no-one-else-will-dare-mention

11 M. Silva, Why is climate 'doomism' going viral – and who's fighting it? *BBC News*, 21 May 2022, https://www.bbc.co.uk/news/blogs-trending-61495035

12 For a good overview of this idea, see: J. Haidt, *The Righteous Mind: Why Good People Are Divided by*

Politics and Religion (London: Penguin, 2013). Also, for evidence of how this plays out in American political life, see: L. Mason, *Uncivil Agreement: How Politics Became Identity* (Chicago: University of Chicago Press, 2018).

13 R. Behr, How Twitter poisoned politics. *Prospect Magazine*, October 2018, pp. 24, 26, https://www.prospectmagazine.co.uk/magazine/how-twitter-poisoned-politics

14 M. Mann, *The New Climate War: The Fight to Take Back Our Planet* (New York: Public Affairs, 2021). Part of what follows is based on my review of this book in the Spring 2021 issue of *Issues in Science and Technology*, pp. 89–90.

15 See Oreskes and Conway, *Merchants of Doubt*.

16 For a general overview of the power of stories, see: S. Dillon and C. Craig, *Storylistening: Narrative Evidence and Public Reasoning* (Abingdon: Routledge, 2021). For an example of how this applies to the case of climate change, see: R. Lejano and S. Nero, *The Power of Narrative: Climate Skepticism and the Deconstruction of Science* (Oxford: Oxford University Press, 2020).

17 M. Shellenberger, Climate change is no catastrophe. Attempts to stop warming will backfire dangerously. *Unherd*, 3 November 2021, https://unherd.com/2021/11/climate-change-will-not-be-catastrophic

18 Organized denial is well covered by Oreskes and Conway in *Merchants of Doubt*. But organized denial is different from other forms of climate change contrarianism or scepticism, which have broader cultural and psychological roots.

5 Why is Climatism Dangerous?

1 T. Knudson, The cost of the biofuel boom: destroying Indonesia's forests. *Yale Environment 360*, 19 January 2009, https://e360.yale.edu/features/the_cost_of_the_biofuel_boom_destroying_indonesias_forests
2 Transport & Environment, 10 years of EU's failed biofuels policy has wiped out forests the size of the Netherlands, 2 July 2021, https://www.transportenvironment.org/discover/10-years-of-eus-failed-biofuels-policy-has-wiped-out-forests-the-size-of-the-netherlands-study
3 Quoted in: G. Ferrett, Biofuels 'crime against humanity'. *BBC News*, 27 October 2007, news.bbc.co.uk/1/hi/world/americas/7065061.stm
4 H.K. Jeswani, A. Chilvers and A. Azapagic, Environmental sustainability of biofuels: a review. *Proceedings of the Royal Society A* 476, 2020: 20200351. This large-scale synthesis of many studies concludes: 'As the findings in this review demonstrate clearly, there are no definitive answers. Even focusing only on the [beneficial climate effects] of biofuels – one of the main drivers for their development – brings with it a host of uncertainties.'
5 J.C. Scott, *Seeing Like a State: How Certain Schemes to Improve the Human Condition Have Failed* (New Haven, CT: Yale University Press, 1998).
6 Louise Amoore powerfully extends Scott's argument across a range of contemporary issues in her book, *The Politics of Possibility: Risk and Security beyond Probability* (Durham, NC: Duke University Press, 2013).
7 For example, see T.J. Lark et al., Environmental outcomes of the US Renewable Fuel Standard. *Proceedings of the National Academy of Sciences* 119(9), 2022: e2101084111.

8 J. Selby, On blaming climate change for the Syrian civil war. *Middle East Report* 296, Fall 2020. See also Marwa Daoudy's account of the war in *The Origins of the Syrian Conflict*.

9 These and other cases are explored in: J.E. Selby, G. Daoust and C. Hoffmann, *Divided Environments: A Political Ecology of Climate Change, Water and Security* (Cambridge: Cambridge University Press, 2022).

10 L. Schipper, V. Castan Broto and W. Chow, Five key points in the IPCC report on climate change impacts and adaptation. *The Conversation*, 3 March 2022, https://theconversation.com/five-key-points-in-the-ipcc-report-on-climate-change-impacts-and-adaptation-178195

11 There are many ticking climate clocks, but this one is particularly prominent. It commenced in September 2019. See: https://climateclock.world/pte

12 UN chief urges leaders of every country to declare 'climate emergency'. *Reuters*, 12 December 2021, https://www.reuters.com/article/uk-climate-change-un-summit-idUSKBN28M0IR

13 This is the condition that critical geographer Eric Swyngedouw and others refer to as 'the post-political'. See for example: E. Swyngedouw, Apocalypse forever? Post-political populism and the spectre of climate change. *Theory, Culture and Society* 27(2–3), 2010: 213–32.

14 M. Hulme, Is it too late (to stop dangerous climate change)? An editorial. *WIREs Climate Change* 11(1), 2020: e630.

15 C. Hickman et al., Climate anxiety in children and young people and their beliefs about government responses to climate change: a global survey. *The Lancet: Planetary Health* 5(12), 2021: e863–73. This

study was publicized by the BBC in September 2021 under the headline 'Climate change: young people very worried – survey', https://www.bbc.com/news/world-5854937

16 For the post-political, see Swyngedouw, Apocalypse forever?; also A. Kenis and E. Mathijs, Climate change and post-politics: repoliticizing the present by imagining the future? *Geoforum* 52, 2014: 148–56; and Y. Pepermans and P. Maeseele, The politicization of climate change: problem or solution? *WIREs Climate Change* 6(4), 2016: 478–85.

17 Although even here there is a very difficult line to draw – and to police – between, on the one hand, constructive, and sometimes penetrating and valid, if awkward, criticism of science and scientists and, on the other, frivolous, misinformed or subversive critiques of scientific claims. For a discussion and analysis see: K.M. Treen, H.T.P. Williams and S.J. O'Neill, Online misinformation about climate change. *WIREs Climate Change* 11(5), 2022: e665.

18 N. Oreskes, There is a new form of climate denialism to look out for – so don't celebrate yet. *Guardian*, 16 December 2015, https://www.theguardian.com/commentisfree/2015/dec/16/new-form-climate-denialism-dont-celebrate-yet-cop-21

19 Freeden, *Ideology*, p. 126.

20 R. Pielke Jr, Stuck between climate doom and denial. *The New Atlantis*, Summer 2022, https://www.thenewatlantis.com/publications/stuck-between-doom-and-denial

21 C. McKinnon, Climate crimes must be brought to justice. *The UNESCO Courier* 19(3), 2019, https://en.unesco.org/courier/2019-3/climate-crimes-must-be-brought-justice

22 The idea of 'wicked problems' has been around for

fifty years, since originally proposed by Horst Rittel and Melvin Webber in their article: Dilemmas in a general theory of planning. *Policy Sciences* 4, 1973: 155–69. Wicked problems are essentially unique, have no definitive formulation and can be considered a symptom of yet other problems. Solutions to wicked problems are difficult to recognize because of complex interdependencies in the system affected, a solution to one aspect of a wicked problem often revealing or creating other, even more complex, problems demanding further solutions. Solving wicked problems is beyond the reach of mere technical knowledge and traditional forms of governance.

23 For both stories, see: R. Bryce, The iron law of electricity strikes again: Germany re-opens five lignite-fired power plants. *Forbes*, 28 October 2022, https://www.forbes.com/sites/robertbryce/2022/10/28/the-iron-law-of-electricity-strikes-again-germany-re-opens-five-lignite-fired-power-plants/?sh=57c3e32f3d0c

24 S. Cornell and A. Gupta, Is climate change the most important challenge of our times? In: M. Hulme (ed.), *Contemporary Climate Change Debates: A Student Primer* (Abingdon: Routledge, 2020), p. 17. Sarah Cornell and Aarti Gupta in their respective essays take different sides on this question and put forward arguments why climate change should or should not be so regarded.

25 The definition of mal-adaptation is from: J. Barnett and S. J. O'Neill, Maladaptation. *Global Environmental Change* 20, 2010: 211–13. The examples, and a broader discussion about the dangers of mal-adaptation, can be found in: S. Juhola, E. Glaas, B-O. Linnér and T-S. Neset, Redefining maladaptation. *Environmental Science & Policy* 55(1), 2016: 135–40.

6 If Not Climatism, Then What?

1 V. Ramachandran, Blanket bans on fossil fuels hurt women. *Nature* 607, 2022: 9. Vijaya Ramachandran is an economist whose research focuses on economic growth and energy infrastructure, mostly in sub-Saharan Africa.

2 R. Atkinson et al., *Climate Pragmatism: Innovation, Resilience and No Regrets* (Oakland, CA: Breakthrough Institute, 2021), https://thebreakthrough.org/articles/climate-pragmatism-innovation It should be noted that in many respects pragmatism is itself an ideology; it also reflects a particular worldview and a political position, and one that I am more sympathetic to. This confirms my earlier argument that we cannot live without ideologies.

3 R. Saravanan's book *The Climate Demon: Past, Present, and Future of Climate Prediction* (Cambridge: Cambridge University Press, 2021) makes this point very clearly. There is also the phenomenon known as 'the certainty trough', which describes how those not closely involved in knowledge production attach much greater certainty to that knowledge than those actually producing it. This was first identified by the sociologist Donald MacKenzie (see for example: Chapter 15, 'The certainty trough' in: R. Williams et al. (eds.), *Exploring Expertise* (Basingstoke: Palgrave, 1998)) and applied to climate modelling by Myanna Lahsen in: Seductive simulations? Uncertainty distribution around climate models. *Social Studies of Science* 35, 2005: 895–922.

4 European Commission, *Destination Earth*, https://digital-strategy.ec.europa.eu/en/policies/destination-earth

5 S. Jasanoff, Technologies of humility. *Nature* 450, 2007: 33.

6 T. Nordhaus, A response to Ezra Klein on 'The

empty radicalism of the climate apocalypse'. *The Breakthrough Institute*, 16 July 2021, https://thebreakthrough.org/blog/a-response-to-ezra-klein-1

7 For a good discussion of the need for caution in thinking about social tipping points, whether they exist and whether they can be controlled, see: M. Milkoreit, Social tipping points everywhere? – Patterns and risks of overuse. *WIREs Climate Change*, 17 November 2022, https://doi.org/10.1002/wcc.813

8 A. Vaughan, UK government admits its net-zero climate strategy doesn't add up. *New Scientist*, 10 June 2022.

9 For a discussion of the ways in which subjective values and expert judgement inflect the notion of climate tipping points, see V. Lam and M.M. Majszak, Climate tipping points and expert judgment. *WIREs Climate Change* 13(6), 2022: e805.

10 M. Roser, The world is awful. The world is much better. The world can be much better. *Our World in Data*, 20 July 2022, https://ourworldindata.org/much-better-awful-can-be-better

11 For the England and Wales figures, see: A. Gasparrini et al., Small-area assessment of temperature-related mortality risks in England and Wales: a case time series analysis. *Lancet Planet Health* 6, 2022: e557–64.

12 T. Nordhaus, Am I the mass murderer? Pushing back on climate catastrophism is not a thought crime. *The Breakthrough Journal*, 29 March 2022, https://thebreakthrough.org/journal/no-16-spring-2022/am-i-the-mass-murderer

13 I explored many of these diverging values and judgements in an earlier book: *Why We Disagree about Climate Change* (2009).

14 S. Bushell, T. Colley and M. Workman, A unified

narrative for climate change. *Nature Climate Change* 5(11), 2015: 971–3.

15 See: M. Verweij, *Clumsy Solutions for a Wicked World* (London: Palgrave Macmillan, 2011). Also useful is the short entry 'Wicked environmental problems', by Michael Thompson in: N. Castree, M. Hulme and J.D. Proctor (eds.), *Companion to Environmental Studies* (Abingdon: Routledge, 2018), pp. 258–62. For a recent review of clumsy solutions and climate change, see: M. Verweij, Clumsy solutions and climate change: a retrospective. *WIREs Climate Change*, 14(1), 2023: e804.

16 Some of the ideas and examples in this section are drawn from my essay 'Climate emergency politics is dangerous', *Issues in Science and Technology* 36, 2019: 23–5.

17 G. J. Wells et al., Tree harvesting is not the same as deforestation. *Nature Climate Change* 12(4), 2022: 307–9.

18 R. Pielke Jr, Welcome to post-apocalyptic climate policy. *The Honest Broker*, 19 April 2021, https://rogerpielkejr.substack.com/p/welcome-to-post-apocalyptic-climate In this essay, Pielke uses the example of population policy as a precedent for such a reframing. The perceived global 'population crisis' of the 1960s and 1970s was transformed from an issue focused on 'overpopulation' to one more focused on seemingly oblique issues, such as women's rights, education, agricultural productivity, democracy, and so on. Issues related to population remain crucially important, but the earlier single focus on 'controlling population' no longer holds sway.

19 Y. Osinbajo, The divestment delusion: why banning fossil fuel investments would crush Africa. *Foreign Affairs*, 31 August 2021, https://www.foreign

affairs.com/articles/africa/2021-08-31/divestment-delusion

20 C. Farbotko, Meta-crisis in a meta-system: does addressing climate change mean new systemic dangers for the world's poor? *WIREs Climate Change* 11(1), 2020: e609.

7 Some Objections

1 W.F. Lamb et al., Discourses of climate delay. *Global Sustainability* 3, 2020: e17, 1–5. All quotations in this paragraph and the next are from this article.

2 S. Shapin, *Never Pure: Historical Studies of Science as If It Was Produced by People with Bodies, Situated in Time, Space, Culture, and Society, and Struggling for Credibility and Authority* (Baltimore, MD: John Hopkins University Press, 2010).

3 Brysse et al., Climate change prediction.

4 For Guterres see: UN Secretary-General's remarks to the Security Council conference on corruption and conflict. New York, 10 September 2018, https://www.un.org/sg/en/content/sg/statement/2018-09-10/secretary-generals-remarks-climate-change-delivered For Bendell see: J. Bendell, Hope and vision in the face of collapse – the 4th R of deep adaptation, 9 January 2019, https://jembendell.com/2019/01/09/hope-and-vision-in-the-face-of-collapse-the-4th-r-of-deep-adaptation For Engelhardt, see: T. Engelhardt, Suicide watch on planet Earth. *Le Monde diplomatique*, 29 April 2019, https://mondediplo.com/openpage/suicide-watch-on-planet-earth See also: B. McGuire, *Hothouse Earth: An Inhabitant's Guide* (London: Icon Books, 2002).

5 Great Barrier Reef Foundation, What is coral adaptation? 17 June 2022, https://www.barrierreef.org/news/explainers/what-is-coral-adaptation

6 E. Cripps, *What Climate Justice Means and Why We*

Should Care (London: Bloomsbury continuum, 2022). See also, for example, Jenny Stephens' criticism of what she calls 'climate isolationism'; 'climate change is not the problem' she argues. See: J.C. Stephens, Beyond climate isolationism: a necessary shift for climate justice. *Current Climate Change Reports* 8, 2022: 83–90.

7 L. Goering, Clean energy transition must be fast and fair, IPCC scientists say. *Reuters*, 4 April, 2022, https://www.reuters.com/article/climate-change-ipcc-society-idUKL5N2W10E3

8 Stephens, Beyond climate isolationism.

9 M. Nisbet, The science of science communication. The March for Science. *Skeptical Inquirer* 41(4), 2017, https://skepticalinquirer.org/2017/07/the-march-for-science

10 See Milton Horowitz's report on the annual meeting of the American Educational Research Association held on 19–21 February 1962, in *Journal of Medical Education* 37, 1962: 634–7.

11 A.H. Maslow, *Psychology of Science: A Reconnaissance* (Washington DC: Henry Regnery, 1969).

12 G.M. Kondolf et al., Save the Mekong Delta from drowning. *Science* 376, 2022: 583–5.

13 Quoted in H.P. Huntington et al., Climate change in context: putting people first in the Arctic. *Regional Environmental Change* 19, 2019: 1217–23.

14 Huntington et al., Climate change in context.

Index

Index

Index

Index

Index

Index

Index

Index

Index

Index

Index